Applied Statistics (Continued)

HANSEN, HURWITZ, and MADOW · Sample Survey Methods and Theory, Volume I
HOEL · Elementary Statistics
KEMPTHORNE · An Introduction to Genetic Statistics
MEYER · Symposium on Monte Carlo Methods
MUDGETT · Index Numbers
RICE · Control Charts
ROMIG · 50–100 Binomial Tables
SARHAN and GREENBERG · Contributions to Order Statistics
TIPPETT · Technological Applications of Statistics
WILLIAMS · Regression Analysis
WOLD and JURÉEN · Demand Analysis
YOUDEN · Statistical Methods for Chemists

Books of Related Interest

ALLEN and ELY · International Trade Statistics
ARLEY and BUCH · Introduction to the Theory of Probability and Statistics
CHERNOFF and MOSES · Elementary Decision Theory
HAUSER and LEONARD · Government Statistics for Business Use, Second Edition
STEPHAN and McCARTHY · Sampling Opinions—An Analysis of Survey Procedures

HARALD CRAMÉR

THE ELEMENTS
OF PROBABILITY THEORY

AND SOME OF ITS
APPLICATIONS

A WILEY PUBLICATION
IN MATHEMATICAL STATISTICS

THE ELEMENTS

OF PROBABILITY THEORY

AND SOME OF ITS APPLICATIONS

By

HARALD CRAMÉR

**PROFESSOR OF ACTUARIAL MATHEMATICS
AND MATHEMATICAL STATISTICS,
UNIVERSITY OF STOCKHOLM**

JOHN WILEY & SONS, NEW YORK

ALMQVIST & WIKSELL, STOCKHOLM

PREFACE

This book is a revised and extended translation of a Swedish textbook which was published first in 1926 and then in entirely rewritten form in 1949.

Starting with a historical introduction to the subject, the book covers the elements of the mathematical theory of probability, with the main emphasis on the theory of random variables and probability distributions. Applications to various fields, particularly to modern statistical methods, are discussed and illustrated by a number of examples. The problems offered for the reader's solution include simple exercises as well as important complements to the theories and methods given in the text.

The book is essentially an *elementary* treatise, and does not aim at a complete and rigorous mathematical development of the subject from an axiomatic point of view. In this respect it can only serve as an introduction to more advanced treatises, such as FELLER's *Probability Theory*, or the present author's *Random Variables and Probability Distributions* and *Mathematical Methods of Statistics*. Occasionally, for the complete proof of some theorem, reference will be made to one of the two last-mentioned works, which will be briefly quoted as *Random Variables* and *Mathematical Methods*.

As for the applications of mathematical probability, any book of moderate size must be content with offering a small sample of problems and methods from the immense field of existing applications, which is continually expanding. Any selection of such a sample by an individual author will necessarily be influenced by a strong personal bias, and most certainly this rule will be found to apply here. Although Part III has been named "Applications", a number of important applications will also be found among the examples and problems that illustrate the theoretical developments in Parts I and II.

It will be assumed that the reader has some working knowledge of analytic geometry, calculus, and algebra, including determinants. Probably the book could be read without difficulty by a student at the junior level in an average American college. Some parts are, of course, more elementary than the others: in particular the first part, which could no doubt be read at an even earlier stage.

My sincere thanks are due to Professor EDWIN HEWITT of the University of Washington, who kindly translated a part of Chapter 16, and to Fil.

5

Lic. Gunnar Blom, who supplied statistical material for some of the examples. I am further indebted to Professor R. A. Fisher and to Messrs. Oliver and Boyd for permission to reprint the tables of the t and χ^2 distributions from *Statistical Methods for Research Workers*, to Professor George W. Snedecor and to the Iowa State College Press for permission to reprint the tables of the F distribution from *Statistical Methods*, and finally to the Princeton University Press for permission to use some of the diagrams from my own *Mathematical Methods*.

University of Stockholm, September 1954.

H. C.

CONTENTS

Part I

FOUNDATIONS

Part II

RANDOM VARIABLES AND PROBABILITY DISTRIBUTIONS

Part III

APPLICATIONS

PART I

FOUNDATIONS

CHAPTER 1

HISTORICAL INTRODUCTION

1.1. Origin of mathematical probability theory. — In order to acquire a real understanding of any branch of science it is necessary to study its historical development. All those problems for which ready-made solutions are offered in modern textbooks were once as many unsolved difficulties facing the scientists of bygone times. The long struggle of past generations against all these difficulties, which has finally led up to our modern science, is certainly worth studying, and the labor spent on such a study will be amply rewarded by the insight into the organic growth and structure of science thus obtained.

The particular branch of science with which we are going to deal in this book is no exception to the general rule. In fact, current literature on mathematical probability theory shows many features that can be fully understood only by a reader who knows something about the historic development of this theory. In this introductory chapter we shall give a very brief sketch of the most essential stages of this development. For a more complete account of the origin and early history of probability theory, the reader may be referred to the classical work by TODHUNTER (see list of references p. 265).

The theory of probability, which at the present day is an important branch of pure mathematics, with a field of applications extending over practically all branches of natural, technical, and social science, has developed from a very humble beginning. Its roots lie in a simple mathematical *theory of games of chance* which was founded about three centuries ago.

In the French society of the 1650's, gambling was a popular and fashionable habit, apparently not too much restricted by law. As ever more complicated games with cards, dice, etc., were introduced, and considerable sums of money were at stake in gambling establishments, the need was felt for a rational method for calculating the chances of gamblers in various games. A passionate gambler, the chevalier DE MÉRÉ, had the idea of consulting the famous mathematician and philosopher BLAISE PASCAL in Paris on some questions connected with certain games of chance, and this gave rise to a correspondence between PASCAL and some of his mathematical friends, above all PIERRE FERMAT in Toulouse.

11

This correspondence forms the origin of modern probability theory. During the rest of the seventeenth century, the questions raised by DE MÉRÉ and further questions of the same kind were discussed among mathematicians. At this early phase of the development, no connected theory of probability had yet been worked out, and the whole subject consisted of a collection of isolated problems concerning various games. In later chapters of this book we shall make detailed acquaintance with these problems. In the present section we shall only make some remarks concerning their general nature.

In all current games of chance with dice, cards, roulettes, and other such apparatus, every single performance of the game must lead to one of a definite number of possible outcomes, represented by the six sides of the die, the 37 cases of the roulette, the 52 cards in an ordinary set of cards, etc. If the gambling apparatus is properly made, and the game is correctly performed, it is not possible to predict in advance which of these possible outcomes will be realized at a particular performance of the game. We cannot predict whether, at the next toss, the coin will fall heads or tails, and similarly in other cases. In fact, this very impossibility of prediction constitutes the *randomness*, the element of uncertainty and chance in the game. On the other hand, there exists between the various possible outcomes of the game a mutual symmetry, which makes us regard all these outcomes as equivalent from the point of view of gambling. In other words, we consider it *equally favourable* for a gambler to risk his stakes on any one of the possible outcomes.

Suppose that we are given a game where this situation is present. Thus every performance of the game will lead to one of a certain number of outcomes, or *possible cases*, and between these possible cases there exists a mutual symmetry of the kind just indicated. Let c denote the total number of these possible cases. Suppose further that, from the point of view of a certain gambler A, the total number c of possible cases can be divided into a group of *favourable cases*, containing a cases, and another group of *unfavourable cases*, containing the remaining $c - a$ cases. By this we mean that, according to the rules of the game, the occurrence of any of the a favourable cases would imply that A wins the game, while the occurrence of any of the $c - a$ unfavourable cases would imply that he loses. If we are interested in estimating the chances of success of A in this game, it then seems fairly natural to consider the ratio a/c between the number a of favourable cases and the total number c of possible cases, and to regard this ratio as a reasonable measure of the chances of gain of A.

Now this is precisely what the classical authors of our subject did. The main difficulty with which they were concerned consisted in the calculation of the numbers a and c of favourable and possible cases in various actual games. As soon as these numbers were known for a given game, their ratio

$$p = \frac{a}{c}$$

was formed. Gradually this ratio came to be known as the *probability* of the *event* which consisted in a gain for the gambler A. This train of thought led to the famous *classical probability definition*, which runs in the following way: *The probability of the occurrence of a given event is equal to the ratio between the number of cases which are favourable to this event, and the total number of possible cases, provided that all these cases are mutually symmetric.*

Though it was not until considerably later that any explicit formulation of a definition of this type appeared, such a definition was more or less tacitly assumed already by PASCAL, FERMAT, and their contemporaries. According to this definition we should say, e. g., that the probability of throwing heads in one toss with a coin is $\frac{1}{2}$, while the probability of obtaining a six in one throw with an ordinary die is $\frac{1}{6}$, the probability of drawing a heart from an ordinary set of 52 cards is $\frac{13}{52} = \frac{1}{4}$, etc.

1.2. Probability and experience. — Already at an early stage, the large mass of empirical observations accumulated in connection with various games of chance had revealed a general mode of regularity which proved to be of the utmost importance for the further development of our subject.

Consider a given game, in each performance of which there are c possible cases which are mutually symmetric in the sense indicated in the preceding section. If this game is repeated under uniform conditions a large number of times, it then appears that *all the c possible cases will, in the long run, tend the occur equally often.* Thus in the long run each possible case will occur approximately in the proportion $1/c$ of the total number of repetitions.

If, e.g., we make a long series of throws with a coin, we shall find that heads and tails will occur approximately epually often. Similarly, in a long series of throws with an ordinary die each of the six sides will occur in approximately $\frac{1}{6}$ of the total number of throws, etc.

If we accept this regularity as an empirical fact, we can draw an important conclusion. Suppose that our gambler A takes part in a game where, in each performance, there are c possible and mutually symmetric cases among which a are favourable to A. Let the game be repeated n times under uniform conditions, and suppose that A wins f times and loses the remaining $n-f$ times. The number f will then be called the *absolute frequency*, or simply the *frequency*, of the event which consists in a gain for A, while the ratio f/n will be denoted as the corresponding *relative frequency* or *frequency ratio*.

Now, if n is a large number, it follows from our fundamental empirical proposition that each of the c possible cases will occur approximately n/c times in the course of the whole series of n repetitions of the game. Since a among these cases are favourable to A, the total number f of his gains should then be approximately equal to an/c. We should thus have, *approximately*, $f = an/c$, or

$$\frac{f}{n} = \frac{a}{c} = p.$$

It thus follows that, according to our empirical proposition, the frequency ratio f/n of A's gains in a long series of games will be approximately equal to the probability p that A wins a game, calculated according to the classical probability definition given in the preceding section.

With a slightly more general formulation, we may express this result by saying that *in the long run any event will tend to occur with a relative frequency approximately equal to the probability of the event.*

Like the classical probability definition, this general principle was not explicitly formulated until a more advanced stage had been reached; already at the time of the chevalier DE MÉRÉ it seems, however, to have been tacitly assumed as an obvious fundamental proposition. As we shall see later (cf. 4.1), one of the questions of DE MÉRÉ was, in fact, directly attached to an application of this general principle to a particular case. In a certain game of chance, DE MÉRÉ had found a lack of agreement between the actually observed frequency ratios of his gains and the value of the corresponding probability of a gain according to his own calculation. It was in order to have this apparent contradiction explained that he consulted PASCAL. However, PASCAL and FERMAT were soon able to show that DE MÉRÉ's calculation of the probability was wrong, and that the correctly calculated probability agreed well with the actually observed frequency ratios, so that no contradiction existed.

1.3. Defects of the classical definition. — The main difficulties encountered at this early stage of probability theory belong to the domain of combinatorial analysis. Starting from certain elementary cases, which are assumed to be completely symmetric — the six sides of the die, the 52 cards of the set, etc. — it is required to combine these according to the rules of some given game so as to form the cases which are possible in that game, always conserving the symmetry between the cases. As soon as we leave the very simplest types of games, this may be a rather intricate task, and it will accordingly often be found that even persons with a good logical training may easily be led into error when trying to solve problems of this kind. Thus it is by no means surprising to find in the early history of our subject a considerable diversity of opinion with respect to the correct way of forming the possible and favourable cases.

As an example of the controversies which occurred in this connection, let us consider a simple game which is closely related to one of the questions of DE MÉRÉ, and which we shall later (4.3) discuss in a more general form.

A and B play heads and tails with a coin which is assumed to be perfectly symmetric, so that each of the two possible outcomes of any throw has a probability equal to $\frac{1}{2}$. The game consists in two throws with the coin. If heads appears in at least one of these throws, A wins, while in the opposite case B wins. What is the probability that A wins the game?

If we denote heads by 1 and tails by 0, the two throws must give one, and only one, of the following four results:

$$00, \quad 01, \quad 10, \quad 11.$$

FERMAT considered these four possible cases as mutually symmetric or, in other words, *equally possible*. Since all these cases, except the first, are favourable to A, he concluded that the probability of A's winning the game is $\frac{3}{4}$. The same result was obtained in a different way by PASCAL.

However, another contemporary mathematician, ROBERVAL, objected that, in the two cases denoted by 10 and 11, A has already won after the first throw, so that it would not be necessary to throw again. Consequently he would recognize only the three possible cases

$$00, \quad 01, \quad 1,$$

and since the last two of these are favourable to A, the probability of A's winning would according to ROBERVAL amount to $\frac{2}{3}$ instead of $\frac{3}{4}$.

Similar objections against the generally accepted rules of probability theory were advanced at a somewhat later stage by D'ALEMBERT. However, it seems fairly obvious that in this way we should lose the perfect symmetry between the cases which is characteristic of the FERMAT solution.

Controversies of this type show that the classical probability definition cannot be considered satisfactory, as it does not provide any criterion for deciding when, in a given game, the various possible cases may be re-regarded as symmetric or equally possible. However, for a long time this defect of the definition was not observed, and it was not until much later that the question was brought under serious discussion.

1.4. Generalization of the probability concept. — About the year 1700 a period of rapid development begins for the theory of probability. About this time two fundamental works on the subject appeared, written respectively by JAMES BERNOULLI and ABRAHAM DE MOIVRE.

The former, one of the members of the famous Swiss mathematical family of the BERNOULLIS, wrote a book with the title *Ars conjectandi* (The Art of Conjecture), which was published in 1713, some years after the death of its author. In this work we find among other things the important proposition known as the BERNOULLI theorem, by which for the first time the theory of probability was raised from the elementary level of solutions of particular problems to a result of general importance. This theorem, which will be fully proved and discussed in chapter 6 of the present book, supplies the mathematical background of those regularity properties of certain frequency ratios in a long series of repetitions of a given game which were indicated above in 1.2.

DE MOIVRE was a French Huguenot who on account of his religion had left France and lived as a refugee in England. His work, *The Doctrine of Chances*, with the subtitle *A method of calculating the probabilities of events in play*, appeared in three editions, 1718, 1738, and 1756, which shows the general interest attracted by our subject during this time. Among other things we find here the first statement of the general theorem known as the *multiplication rule* of the theory of probability, which will be proved below in 3.3. In the last two editions of this remarkable book we also find the first indications of the *normal probability distribution* (cf. chapter 7), which at a later stage was to play a very important part in the development of the subject.

In the works of BERNOULLI and DE MOIVRE, the theory of games of chance was further developed on the basis of the (more or less tacitly used) classical probability definition, and various combinatorial and other mathematical methods were applied to the theory. It is a characteristic feature of the early history of probability theory that there was at this

16

stage a very intimate contact between the development of that particular theory and the general mathematical development. Later on, this contact was to a large extent lost, and it is only fairly recently that a fruitful contact has again been established.

About this time, a new and extremely important idea appears. It was found that the terminology and the rules of calculation of probability theory, which had been introduced with the sole intention of building up a mathematical theory of games of chance, could be applied with good results also to various problems of entirely different types, some of which fall clearly outside the scope of the classical probability definition.

Such was the case, e.g., in the statistics of human populations and in the mathematical theory of life insurance, two closely allied fields which were both in a state of vigorous development during the eighteenth century. As an example of the applications encountered in these fields, suppose first that we observe the sex of each of a series of new-born children. We can then formally regard our observations as a sequence of repetitions of a game of chance, where in each repetition we have the two possible outcomes "boy" and "girl". Similarly, if we observe a certain number of persons of a given age during a period of, say, one year, and note for each person whether he is alive at the end of the year or not, we can also in this case talk of a game of chance with the two possible outcomes "life" and "death". By statistical experience it was soon found that the frequency ratios which can be formed for observations of these kinds, such as the frequency ratio for the outcome "boy" in a sequence of observations on new-born children, show a similar tendency to become more or less constant in the long run, as we have already pointed out for frequency ratios connected with ordinary games of chance (cf. 1.2). This being so, it seemed quite natural to talk of "the probability that a new-born child will be a boy" or "the probability that a man aged 30 years will die within a year", in just the same way as we talk of "the probability of obtaining a six in a throw with an ordinary die". Moreover, it seemed quite as natural to apply to these new probabilities the whole terminology and the rules of calculation worked out in the theory of games of chance. In this way, it soon proved possible to reach without difficulty a large number of interesting and practically useful new results, such as methods for the calculation of life tables, annuity values, life insurance premiums, etc.

However, during this extension of the domain of application of the theory, due attention was not paid to the fundamental question of the basic probability definition. According to the classical definition, mathe-

17

matical probability is defined in terms of the numbers of certain possible and favourable cases, which are assumed to be mutually symmetric. It remained an unsolved, and practically untouched, question how this definition should be interpreted and how the possible and favourable cases should be conceived outside the domain of games of chance, e.g., in connection with probabilities of events such as "boy" or "death" in the examples considered above.

The result of this process of extension, which took place during the eighteenth century, was summed up in a very interesting form in the works of the famous LAPLACE, and particularly in his classical treatise *Théorie analytique des probabilités*, first published in 1812. This work contains in the first place a systematic and very complete account of the mathematical theory of games of chance. In addition, it contains a large number of applications of probability theory to a great variety of scientific and practical questions. For these investigations, Laplace uses throughout the most modern tools of the mathematical technique of the period. In an extensive introduction written for non-mathematical readers, he explains his general views on all these questions, and his appreciation of the value of the results reached by the aid of probability theory. The study of this introduction, which is a typical specimen of the optimistic rationalism of the time, can be highly recommended to anybody interested in the history of science.

With respect to questions of basic definitions, however, Laplace takes a very uncritical attitude. He seems, in fact, to hold that the classical probability definition is everywhere directly applicable. Without apparently feeling the need for any further investigation, he thus regards any of his applications as directly comparable to a game of chance, where the possible outcomes naturally divide into a number of mutually symmetric cases. However, we are left without any indication as to how this division should be performed in an actual case falling outside the domain of the games of chance.

1.5. Applications. — The work of Laplace exercised a profound influence on the subsequent development of the subject. In view of the impressive mathematical apparatus and the important practical results already obtained or within easy reach, it was tempting to overlook the fatal weakness of the conceptual foundations. As a result, the field of applications of probability theory expanded rapidly and continually during the whole of the nineteenth century, while the mathematical probability theory itself showed during the same time a marked tendency to stagnation.

We shall briefly mention some of the most important types of applications originating from this period. GAUSS and LAPLACE discussed independently of each other the applications of mathematical probability theory to the numerical analysis of errors of measurement in physical and astronomical observations. They largely improved earlier attempts in this direction, and worked out a systematical *theory of errors*. This theory and the closely related *method of least squares* have become of great practical and theoretical importance.

The enormous development of life insurance since the beginning of the nineteenth century was rendered possible by a corresponding development of *actuarial mathematics*, which in its turn is based on the application of probability to *mortality statistics*. Further applications to *demography* and other branches of social science were made by QUETELET and his school.

In a quite different field, viz., mathematical physics, probability theory was introduced by the work of MAXWELL, BOLTZMANN, and GIBBS on *statistical mechanics*, which has been of fundamental importance for great parts of modern physical science.

During the present century, this development has continued at an accelerated pace. Methods of *mathematical statistics* have been introduced into a steadily growing number of fields of practical and scientific activity. The mathematical theory on which these methods are based rests essentially on the foundation of mathematical probability. At the present moment, the applications of probability theory cover such widely different fields as *genetics*, *economics*, *psychology*, and *engineering*, to mention only a few randomly chosen examples in addition to those discussed above.

1.6. Critical revision of foundations. — As already indicated in the preceding section, the rapid development of the applications of probability theory during the period immediately following the publication of the fundamental work of LAPLACE was accompanied by a stagnation in the mathematical part of the subject. The earlier intimate contact between probability theory and general mathematical analysis was gradually lost, and even during the early part of the present century our subject remained almost unaffected by the demand for logical precision and rigour that had firmly established itself in most other branches of mathematics.

However, now and then critical voices were heard, drawing attention to the strictly limited scope of the classical probability definition, and to the necessity of building the whole structure of probability theory on more stable foundations. Gradually, it became evident that a thoroughgoing critical revision was indispensable.

19

The first attempt to overcome the difficulties consisted in analyzing the concept of *symmetric* or *equally possible* cases used in the classical definition, and trying to improve this definition by including some appropriate criterion for the symmetry or equal possibility of cases. Among the authors whose work follows this line may be mentioned particularly BERTRAND and POINCARÉ.

However, in other quarters it was strongly held that, in many of the most important practical applications of probability, it is difficult or even impossible to form any concrete idea at all about the nature of the division into a certain number of equally possible cases required by the classical definition. This difficulty is quite evident even in such a simple example as the previously mentioned probability that a person of given age will die within a year. It is certainly not easy to see how in this example we could think of some equally possible cases corresponding to the six sides of a die.

By arguments of this type, many authors were led to look for a fundamentally new approach to the subject, trying to replace the classical probability definition by some entirely new definition, more or less directly founded on the *stability properties of frequency ratios* briefly indicated in 1.2. The frequency ratio of a certain event in a sequence of observations was then regarded as an observed value of a hypothetical quantity which, by definition, was conceived as the probability of the event.

If the theory is built on a probability definition of this type, the classical definition will obviously lose its position as the basic definition of the theory. It will now simply serve as a convenient *rule for finding the value of a probability* which is applicable as soon as a division into equally possible cases can be performed. Examples of this situation occur, as we shall see later, not only in problems connected with games of chance, but also in certain other cases that are often of great practical importance.

Since about 1850, there have been repeated attempts to introduce a suitable probability definition of the frequency type, and to build the theory on such a definition. During the present century, investigations in this field have been more and more influenced by the *tendency to axiomatization*, which is a characteristic feature of modern mathematics. According to this point of view, the probability of an event is conceived as a numerical quantity associated with that event, and assumed to possess certain basic properties expressed by *axioms*, i.e., fundamental propositions stated and accepted without proof. The probability definition is then given in the following form: *a probability is a numerical quantity satisfying such and such axioms.*

In connection with this recent development, the old intimate contact between probability theory and mathematics in general has been re-established. Modern probability theory is, fundamentally, a branch of pure mathematics, engaged in a fruitful and stimulating exchange of ideas and impulses with other branches. At the same time, the field of applications of the theory is still continually increasing.

The account of probability theory given in this book will be based on a definition of the frequency type. As this is essentially an elementary treatise, we shall not attempt to give a mathematically rigorous and complete axiomatic treatment, which would necessarily presuppose a certain degree of mathematical maturity in the reader. A more advanced account of the subject from this point of view will be found in the works by FELLER, KOLMOGOROFF, and the author of the present book, which are quoted in the list of references. A closely related, though in some respects different, point of view is represented by the works of VON MISES also quoted in the list of references.

CHAPTER 2

DEFINITION OF MATHEMATICAL PROBABILITY

2.1. Mathematical models. — According to that view on the foundations of probability theory on which the present book is based we can say that, generally speaking, the object of probability theory is *to provide a mathematical model, suitable for the description and interpretation of a certain class of observed phenomena.* — Before proceeding to a detailed analysis of this statement, it will be convenient to present in this section a few remarks of a general character, concerning the relations between a group of observed phenomena and a mathematical theory designed for their interpretation.

Any attempt at mathematical description of observed phenomena must involve a certain idealization of really observed facts. The formulae of the mathematician can only provide a simplified *mathematical model* of the world of reality, a kind of idealized picture of the characteristic features of the phenomenon under investigation.

If, e.g., we draw geometrical figures on a black board by means of a piece of chalk, a ruler and a compass, it will be possible by purely empirical measurement to discover and to verify the validity of certain simple rules, which hold more or less accurately for these concrete figures.

21

Thus we shall find in this way that the sum of the angles in a concrete triangle on our black board will be *approximately* equal to 2π, and in a similar way we can verify the *approximate* validity of various further geometrical theorems for our concrete figures.

In order to give a simple mathematical description of these empirically found regularities, it is well known that ordinary euclidean geometry introduces certain purely theoretical concepts, such as mathematical points, lines and figures, which can be conceived as idealized counterparts of the concrete figures on our black board. The fundamental properties of these concepts are postulated by the geometrical axioms, and from these all the propositions of euclidean geometry are then obtained by purely logical deduction. Within the mathematical model constructed in this way, the ordinary geometrical propositions hold not only approximately, but *exactly*, with logical necessity. Thus in a conceptual mathematical triangle the sum of the angles will be *exactly* equal to 2π, and similarly with respect to the validity of all other propositions of euclidean geometry.

Mathematical models of the same general structure are used, e.g., in theoretical mechanics and in other branches of physics. By experimental measurement of various physical quantities, we can verify certain regularities that hold more or less accurately for the observed values. In a similar way as the geometer, the physicist then introduces an idealized mathematical model, where the experimentally observed values of various physical constants are replaced by hypothetical *true values,* and where the *exact* validity of the physical laws for these hypothetical values is postulated.

Whether such a mathematical theory can be considered as a satisfactory model of the really observed phenomena is a question that can be decided only by experience. The answer will depend on the degree of agreement between the consequences of the mathematical theory and our concrete observations. When, e.g., we say that euclidean geometry provides a mathematical model of reality fully adequate for all ordinary practical purposes, this is a statement which, by the nature of things, can neither be proved nor disproved by mathematical reasoning, but only by experience. It is well known that experience complety confirms this statement.

We shall now see in the sequel that the theory of probability can be considered as a mathematical model of the same kind as those indicated above.

2.2. Random experiments. — In many different fields of practical and scientific activity, we encounter cases where certain experiments or observations may be repeated several times under uniform conditions, each individual observation giving a certain definite result.

The games of chance discussed in the preceding chapter provide a simple example of experiments of this type, but there are plenty of other and more important examples. Thus in the usual type of scientific experiment occurring in physics, chemistry, biology, medicine, etc., it is endeavoured to keep all relevant conditions of the experiment under control, so as to render a uniform repetition possible. Similarly, in most branches of industry, the production process involves the continual repetition of certain operations under conditions that should be kept as constant as possible. Further examples are found in demographic and social statistics, where we often observe certain properties in a sequence of individuals selected in a uniform way from some given population.

Now in such cases we practically always find that, even though the utmost care is taken to keep the conditions of the experiment as uniform as possible, there will appear an *intrinsic variability* that cannot be kept under control. Owing to this variability, the result of the experiment will vary in an irregular way between successive repetitions, and the result of an individual repetition cannot be accurately predicted.

As already pointed out in 1.1, this is precisely what happens in every game of chance. The outcomes of successive repetitions of a given game will always vary in a random and unpredictable way. But the same kind of random variability appears, in a more or less palpable way, in all the other examples given above. Thus, e.g., successive measurements of some physical or astronomical quantity will as a rule not yield identical results, and we are accustomed to ascribe the variations to "errors of measurement", which is simply a convenient name for a source of variation that eludes control. A similar random and unpredictable variability appears, as a more or less dominating feature, in all kinds of repeated experiments, production processes and statistical observations.

In any such case, we shall say that we are concerned with a random experiment or a random observation.

Thus in any given random experiment E, the result should be expected to vary from one repetition of the experiment to another. In general, however, it will be possible to assign a definite set S consisting of all results which are recognized as a priori possible outcomes of the experiment E. Let us consider some simple examples.

a) If E consists in observing the sex of a newborn child, the set S will consist of the two possible outcomes "boy" and "girl".

23

b) If E consists in a throw with an ordinary die, S will consist of the six numbers 1, 2, . . ., 6.

c) Let E consist in counting the number of telephone calls on a given trunkline during an interval of, say, ten minutes. Since there is no obvious upper limit to this number, the set S should here be taken to consist of all non-negative integral numbers 0, 1, 2, . . .

Let us now consider a well-defined random experiment E, and the corresponding set S of all possible results of E. Consider further the *event* A that the result of the experiment E belongs to some a priori specified part of the set S, e.g., the event that the newborn child is a boy, that the die turns up an odd number of points, that the number of telephone calls exceeds 50, etc. In any performance of E, the event A may then occur, or it may fail to occur.

Suppose that we perform a sequence of n repetitions of the experiment E, trying to keep all relevant conditions as constant as possible. Let f denote the number of those repetitions where the event A occurs, so that in the remaining $n - f$ repetitions A fails to occur. In accordance with the terminology already introduced in 1.2, we shall then call f the *absolute frequency*, or briefly the *frequency*, of the event A in our sequence of n repetitions of E, while the ratio f/n is the corresponding *relative frequency* or *frequency ratio*.

Suppose further that we perform several sequences of repetitions of the same experiment E, so that we obtain several frequency ratios f_1/n_1, f_2/n_2, . . ., for the event A. In many important cases it will then be found that *these frequency ratios only differ from one another by small amounts, as soon as the corresponding numbers of repetitions $n_1, n_2, . . .$ are large.* Thus in a case of this type there is a remarkable inherent *tendency to stability* in the frequency ratios, a tendency to accumulate in the neighbourhood of some fixed value, which reminds of the way in which a series of measurements of some physical constant accumulate in the neighbourhood of the hypothetical true value of that constant.

The same tendency to stability is encountered when, in one single sequence of repetitions of our experiment E, we plot in a diagram the frequency ratio f/n of the event A against the number n of repetitions. For increasing values of n, we shall then get a graph of the general type illustrated in Fig. 1, where the tendency of the frequency ratios to accumulate in the neighbourhood of some value approximately equal to $\frac{1}{2}$ is unmistakable.

In the particular case when the experiment E is a simple game of chance, this property of the frequency ratios has been mentioned already

24

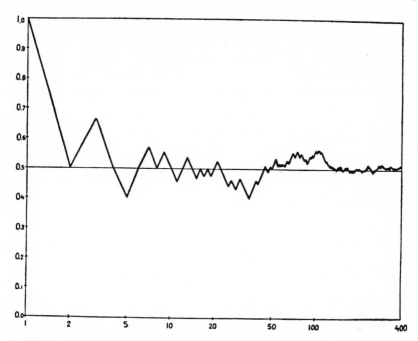

Fig. 1. Frequency ratio for "heads" in a sequence of throws with a coin. Logarithmic scale for the abscissa.

in 1.2, where we have also seen that the value in the neighbourhood of which the frequency ratios will accumulate can in this case be determined by means of the classical probability definition. For our present purpose it is, however, important to note as an empirical fact that the *long run stability of frequency ratios* has been found to apply to a very large class of random experiments, of which games of chance only form a rather insignificant particular case.

If, for a given random experiment E, this mode of stability of frequency ratios appears for various events A considered in connection with the experiment, we shall say that E shows *statistical regularity*, i.e., a regularity of the average results of long sequences of repetitions, as distinct from the random irregularity of the results of individual repetitions. Any systematic record of the results of sequences of repetitions of this kind will be said to constitute a *set of statistical data* relative to the phenomenon concerned.

The object of this book will be to work out a mathematical model for the description and interpretation of phenomena showing statistical regularity, and to show how this model can be used for the drawing of valid inferences

25

from statistical data. The name of this model is the mathematical theory of probability.

The reader should verify for himself the above general statement concerning the stability properties of frequency ratios by means of experiments with coins and dice, by computing some sex ratios for successive months or years from official statistics, etc. The results of such experiments may sometimes be presented by a diagram of the type shown in fig. 1, and sometimes by a simple comparison of the resulting frequency ratios.

An experiment of the last-mentioned type was performed in the following way. In the first place, five series of 10 throws each were made with a coin, and the highest and the lowest of the frequency ratios obtained for heads were noted. The same procedure was then repeated with five series of 100 throws each, and finally with five series of 1000 throws each. The result is given in the following table, which shows how the difference between the extreme frequency ratios decreases as the length of the series increases, and how at the same time the frequency ratios seem to accumulate in the neighbourhood of some value approximately equal to $\frac{1}{2}$.

Number of throws in each series	Frequency ratio Maximum	Frequency ratio Minimum	Difference
10	0.600	0.300	0.300
100	0.550	0.480	0.070
1000	0.507	0.496	0.011

2.3. Frequency ratios and mathematical probabilities. — As in the preceding section we consider a series of n repetitions of the random experiment E, and denote by f/n the corresponding observed frequency ratio of the event A. If our experiment E shows statistical regularity, and if we observe some values of the frequency ratio of A in series with large numbers of repetitions n, we may expect that the observed frequency ratios will only differ among themselves by insignificant amounts.

In order to prepare for a simple mathematical description of the behaviour of the frequency ratios for large values of n, it then seems natural to postulate the existence of a number P which can be conceived as a mathematical idealization of the frequency ratio f/n, in the same way as the hypothetical true value of some physical constant constitutes a mathematical idealization of our empirical measurements, or as a mathematical straight line provides an idealized counterpart of a concrete line on the black board.

This number P will, by definition, be called the mathematical probability of the event A in connection with the random experiment E. The frequency ratio f/n will then be considered as an experimental measurement of the probability, corresponding to the true value P. In accordance with the general empirical proposition concerning the long run stability of frequency ratios we *may expect that, for large values of n, the observed frequency ratio f/n will be approximately equal to the probability P.*

It should be well observed that, according to this definition, the event A and the experiment E will both have to be given before the probability P can be determined. As soon as we are concerned with the value of a probability in some practical application, it will thus always be necessary to take care that both the event A and the experiment E are strictly specified. When it is desirable to use a notation which explicitly shows the dependence of P on both these arguments, we shall write

$$P = P(A, E).$$

However, in the majority of practical applications, the situation is that we are concerned with the mutual relations between the probabilities of various events A attached to a certain random experiment E which is considered as fixed once for all. In such cases we may omit E in the notation, and write

$$P = P(A),$$

which implies that the probability P is then considered as a *function of the event A*.

In the following chapter we shall study some of the simplest properties of this function. Before proceeding to do this let us, however, for one moment consider the important question how the probability P can be determined when A and E are given. At the present stage we must restrict ourselves to some general introductory remarks concerning this question, to which we shall return later (cf. 14.3).

In certain problems, we encounter probabilities that can be determined *a priori*, without the aid of a study of the really observed frequency ratios. As already pointed out in 1.6, this will be possible as soon as the rule provided by the classical probability definition is applicable, i.e., as soon as a division into symmetric or equally possible cases can be performed. Also in certain other problems it may sometimes be possible to determine the value of a probability by means of a priori arguments.

When this cannot be done we must fall back upon a determination *a posteriori*, i.e., by estimating the value of an unknown probability by means of a study of the behaviour of the frequency ratios of the corresponding event. We shall later discuss the methods that can be used for such estimations. At the present moment it will be sufficient to observe that, according to our general probability definition, the logical superiority will always belong to the aposterioric method. If we assert, e.g., that on account of symmetry arguments, the probability of heads at a throw with a certain coin is $\frac{1}{2}$, this value can, for a given concrete coin, only be regarded as approximate. In fact, the symmetry of the coin will never be perfect, and the probability of getting heads at a throw with the coin should be considered as a physical constant of that particular coin. If the coin is well made, the value of this constant may be expected to be approximately equal to $\frac{1}{2}$, but still there is always the possibility that we could obtain a better approximation to the observed frequency ratios by using some value slightly different from $\frac{1}{2}$, which could be estimated by means of the aposterioric method.

We shall conclude this section by some remarks concerning the scope of our probability definition. We have defined the concept of mathematical probability only for an event attached to some random experiment, which can be repeated several times under uniform conditions. Actual repetition may, of course, sometimes be difficult or even impossible to realize, e.g., owing to prohibitive costs of experimentation, but it is sufficient that a repetition under uniform conditions should be *conceivable*. — On the other hand, in everyday language the word probability is often used in a much more general sense that falls entirely outside the scope of our definition. When, e.g., we talk of the probability that SHAKE-PEARE's plays were written by BACON, or the probability that there exists organic life on the planet of Mars, we are concerned with events which do not seem related to any kind of random experiments, so that our definition is not applicable. The question whether the concept of mathematical probability can be given a definition sufficiently wide to include cases of these types, and whether such a definition would serve any useful purpose, has been extensively discussed by a large number of authors. In this book we shall, however. entirely refrain from the discussion of these questions, confining ourselves exclusively to probabilities which, in accordance with the above definition and with the rules for calculation with probabilities that will be laid down in the sequel, can be brought into direct connection with random experiments and frequency ratios.

CHAPTER 3

ELEMENTARY RULES OF CALCULATION FOR PROBABILITIES

3.1. Some simple properties of mathematical probability. — In the preceding chapter we have defined mathematical probability, and we must now make acquaintance with some of the most important properties of this concept. The basic properties of mathematical probabilities will be obtained by an appropriate mathematical idealization of the properties of observed statistical frequency ratios, in the same way as the geometrical axioms are obtained by an idealization of the observed properties of concrete geometrical figures.

Let us consider a random experiment E, the conditions of which are fixed once for all. Let f denote the absolute frequency of the event A in a series of n repetitions of E. It is then obvious that we shall always have $0 \leq f \leq n$, and hence

$$0 \leq \frac{f}{n} \leq 1 .$$

For large values of n, however, the frequency ratio f/n will be an approximate value of the corresponding probability $P(A)$, and thus it is natural to require that the probability $P(A)$ should satisfy the same inequality, so that we should always have

$$0 \leq P(A) \leq 1.$$

Every probability will thus belong to the interval between zero and one, the extreme values included.

If, in particular, A denotes a *certain* event, i.e., an event which will necessarily occur at each performance of the experiment E, we shall always have $f = n$, and thus $f/n = 1$. By the same argument as above, we must then require that the corresponding probability should also be equal to one. *The probability of a certain event is thus always equal to one.*

If, on the other hand, A denotes an *impossible* event, i.e., an event which can never occur at any performance of E, the frequency f and the frequency ratio f/n will both always be equal to zero. In the same way as above it then follows that *the probability of an impossible event is always equal to zero.*

29

Thus, e.g., the probability of obtaining in a throw with an ordinary die one of the numbers $1, 2, \ldots, 6$ is equal to one, since this is a certain event. On the other hand, the probability of obtaining in the same throw the number 7 is equal to zero, since this is an impossible event.

The two last propositions printed in italics cannot in general be converted. Thus if the probability of the event A is known to be equal to zero, it will only be possible to assert that the frequency ratio f/n can be expected to take very small values as soon as the number n of repetitions is large, which may be expressed by saying that A is an *unusual* event. In general, it will not be legitimate to make the stronger assertion that the frequency ratio is strictly equal to zero, i.e., that A is an *impossible* event. — A similar remark holds in respect of an event of probability one.

Even if we only know that the probability of the event A is *small* (although not necessarily equal to zero), it will obviously be possible to conclude that the frequency ratio of A in a long series of repetitions of E can be expected to take very small values, so that A is an unusual event. In the applications of mathematical probability, events of this kind play a great part. A particularly important type of situation where this case occurs is when, starting from certain assumptions, we can show that the probability of a given event is very small. We then conclude that, if our assumptions are true, this event A must be so unusual that we can regard it as *practically certain* that A will not occur, if we make one single or a small number of performances of the experiment E. If, in an actual performance of the experiment, A occurs, this should then be taken as an indication that there may be something wrong about our assumptions. — We shall return to questions of this type in Part III.

3.2. The addition rule. — Let now A and B denote two events, each of which may occur, or fail to occur, at any performance of the given random experiment E. We may then consider the composed event which consists in the occurrence of *at least one* (possibly both) of the events A and B. It seems rather natural to consider this event as the *sum* of the events A and B, and we shall accordingly denote it by $A + B$. Similarly, the event that consists in the occurrence of *both* events A and B will be considered as the *product* of A and B, and will accordingly be denoted by $A B$.[1]

---

[1] A reader acquainted with the elements of the mathematical theory of sets will immediately see that we are here in reality concerned with sums and products (or, in another terminology, unions and intersections) of sets. In a strictly mathematical ac-

We now perform a series of n repetitions of the experiment E. Let f_A, f_B, f_{A+B} and f_{AB} denote the numbers of occurrences of the events A, B, $A+B$ and AB respectively. Obviously the following identity will then hold between these frequencies:

$$f_{A+B} = f_A + f_B - f_{AB}.$$

If every term in this equation is divided by n, we obtain an identity of the same form between the corresponding frequency ratios. By the same argument as in the preceding section, it then follows that we must require that the same relation should hold for the corresponding probabilities. We thus obtain the following formula, which is known as the *addition rule* of probability theory:

(3.2.1) $$P(A+B) = P(A) + P(B) - P(AB).$$

Ex. 1. — One card is drawn from each of two ordinary sets of 52 cards. The probability that at least one of them will be the ace of hearts is then

$$\frac{1}{52} + \frac{1}{52} - \frac{1}{52^2} = \frac{103}{2704}.$$

In fact, denoting by A the event that the first card is the ace of hearts, and by B the corresponding event in respect of the second card, we immediately find $P(A) = P(B) = \frac{1}{52}$. Further, in order to calculate the probability $P(AB)$ that both cards are aces of hearts, we have the 52^2 symmetrical possible cases which are represented by all possible pairs consisting of one card from the first set and one from the second. Only one of these cases is favourable to the event AB, so that $P(AB) = \frac{1}{52^2}$, and 3.2.1 gives the result stated above.

The addition rule is easily extended to the case of more than two events. If in 3.2.1 we replace B by $B+C$, we obtain without difficulty the formula

(3.2.2)
$$\begin{aligned}
P(A+B+C) = {} & P(A) + P(B) + P(C) \\
& - P(AB) - P(AC) - P(BC) \\
& + P(ABC),
\end{aligned}$$

where $A+B+C$ denotes the event which consists in the occurrence of at

---

count of the theory of probability, the vaguely defined concept of an *event* will, in fact, be replaced by the concept of a *set of possible outcomes* of the experiment E. The occurrence of at least one of the events A and B will then correspond to the sum (or union) of the sets of outcomes corresponding to A and B, while the joint occurrence of A and B will correspond to the product (or intersection) of those two sets.

least one of the events A, B and C, while ABC stands for the simultaneous occurrence of all three events.

By a repeated application of the same procedure, the general addition rule for an arbitrary number of events is obtained in the following form, where the notations should be easily understood:

$$
\begin{aligned}
P(A_1 + A_2 + \cdots + A_m) = {} & P(A_1) + P(A_2) + \cdots + P(A_m) \\
& - P(A_1 A_2) - P(A_1 A_3) - \cdots - P(A_{m-1} A_m) \\
& + P(A_1 A_2 A_3) + P(A_1 A_2 A_4) + \cdots + P(A_{m-2} A_{m-1} A_m) \\
& \cdots\cdots\cdots\cdots\cdots\cdots\cdots\cdots\cdots\cdots\cdots\cdots\cdots \\
& (-1)^{m-1} P(A_1 A_2 \cdots A_m).
\end{aligned}
$$

(3.2.3)

There is an important particular case when the addition rule can be written in a considerably simplified form. Two events A and B will be called *mutually exclusive* if they can never occur at the same time. In this case the product event AB is an impossible event, so that by the preceding section we have $P(AB) = 0$, and thus 3.2.1 takes the form

(3.2.1 a) $$P(A + B) = P(A) + P(B).$$

Ex. 2. — Two cards are drawn simultaneously from the same set. The probability that at least one of them will be the ace of hearts is then $\dfrac{1}{52} + \dfrac{1}{52} = \dfrac{1}{26}$. In fact, both cards cannot be the ace of hearts, so that in this example the events A and B are mutually exclusive, and we can use the formula 3.2.1 a.

An important particular case of 3.2.1a is obtained if we denote by A^* the *opposite event* to A, i.e., the event which occurs whenever A fails to occur.[1] The two events A and A^* will then be mutually exclusive, and further $A + A^*$ is a certain event, so that 3.2.1 a gives

(3.2.4) $$P(A) + P(A^*) = 1.$$

The sum of the probabilities of any event and its opposite event is thus always equal to unity. — Instead of directly calculating the probability of a given event A, we may thus first calculate the probability of the opposite event A^*, which will sometimes be simpler, and then subtract this probability from unity, thus obtaining the probability of A. In the sequel we shall often have occasion to use this method of calculation.

---

[1] In the language of the theory of sets, A^* is the set of outcomes of our experiment E which is *complementary* to the set A.

The general addition rule 3.2.3 is simplified in a similar way for the case of mutually exclusive events. If any two among the events $A_1, A_2, \ldots, A_m$ are mutually exclusive, 3.2.3 will give

$$(3.2.3\text{ a}) \qquad P(A_1 + A_2 + \cdots + A_m) = P(A_1) + P(A_2) + \cdots + P(A_m).$$

3.3. Three fundamental propositions. — In the two preceding sections, we have deduced various properties of the probability numbers by means of an idealization of the corresponding properties of observed frequency ratios. It is easily seen that the properties thus obtained are not all logically independent. In fact, all these properties can be deduced from the following three basic properties, which we shall call the fundamental propositions of our theory:

I. *Any probability $P(A)$ is a non-negative number: $P(A) \geqq 0$.*

II. *The probability of a certain event is equal to unity.*

III. *If the events A and B are mutually exclusive, we have the addition rule: $P(A + B) = P(A) + P(B)$.*

These three propositions contain all the properties which it will be necessary to found on an idealization of the empirically known properties of statistical frequency ratios. If the three fundamental propositions are taken as given, it is possible to deduce all properties given in the two preceding sections by purely logical arguments, without any further reference to observed frequency ratios.

We shall now take two examples from among the properties discussed in the two preceding sections, and show how these can be deduced from the fundamental propositions. It will be left to the reader to work out similar deductions for the other properties given in 3.1–3.2.

We first observe that the fundamental proposition I only states that any probability $P(A)$ satisfies the inequality $P(A) \geqq 0$, while in 3.1 we have given the more restrictive inequality $0 \leqq P(A) \leqq 1$. Let us now show how this can be deduced from the fundamental propositions. In fact, since A and A^* are mutually exclusive, and since $A + A^*$ is a certain event, it follows from II and III that

$$P(A) + P(A^*) = P(A + A^*) = 1.$$

However, by I we have $P(A^*) \geqq 0$, and thus it follows that $P(A) \leqq 1$.

As our second example we choose the addition rule 3.2.1 for two arbitrary events, observing that in the fundamental proposition III we have been concerned only with the particular case of two mutually exclusive events, when the addition rule takes the simplified form 3.2.1 a. In order to deduce the addition rule 3.2.1 for two arbitrary events A and B, we first observe that the events A and A^*B are mutually exclusive, and that we obviously have $A + B = A + A^*B$. By III we then have

$$P(A + B) = P(A + A^*B) = P(A) + P(A^*B).$$

Further it is easily seen that the events AB and A^*B are mutually exclusive, and that $B = AB + A^*B$, so that

$$P(B) = P(AB + A^*B) = P(AB) + P(A^*B).$$

By subtraction we then directly obtain 3.2.1.

The three fundamental propositions will be taken as the basis of probability theory as developed in this work. In the two first parts of the book we shall be concerned with the development of the mathematical theory, while in the third part we shall consider this theory as a mathematical model of certain real phenomena, and discuss various applications of this model.

The three fundamental propositions cannot be regarded as a *system of axioms* for probability theory, in the rigorous mathematical sense, as we are here dealing with the concept of an *event*, which has not been given a precise mathematical definition. With the aid of the mathematical theory of sets and measure of sets, the requisite precise definitions can be given, and in this way the fundamental propositions can be worked out to a real system of axioms for probability theory. However, this would surpass the level of mathematical knowledge for which this book has been planned, so that we must content ourselves by referring the reader on this point to the literature which has been indicated in 1.6. — In this connection we shall here only observe that, in recent mathematical literature, the addition formula III is usually stated with an important extension, in the following form: let A_1, A_2, ... be a *finite or infinite* series of events, such that any two among them are mutually exclusive. It is then required that $P(A_1 + A_2 + \cdots) = P(A_1) + P(A_2) + \cdots$.

We shall finally show that the rule for the determination of the value of a probability which is implied by the classical probability definition (cf. 1.1) can easily be deduced from the fundamental propositions. Suppose that there are in all c possible mutually exclusive outcomes of the random experiment E, and let us denote these by $E_1, E_2, \ldots, E_c$. In accordance with the classical definition, we shall assume these to be mutually symmetric or equally possible cases, which is obviously only another way of saying that all probabilities $P(E_1), \ldots, P(E_c)$ are assumed to be equal. As $E_1 + E_2 + \cdots + E_c$ is a certain event, it then follows from III and II that

$$P(E_1) + \cdots + P(E_c) = P(E_1 + \cdots + E_c) = 1,$$

34

and since all the $P(E_i)$ are equal, each of them must have the value $1/c$. Suppose further that a of these outcomes, say E_1, E_2, . . ., E_a, are favourable to the occurrence of some given event A. We then have

$$P(A) = P(E_1 + \cdots + E_a) = P(E_1) + \cdots + P(E_a) = \frac{a}{c},$$

in accordance with the rule of the classical definition.

3.4. Conditional probabilities and the multiplication rule. — Using the same notations as in the deduction of the addition rule in 3.2, we now consider a series of n repetitions of a given random experiment E. From the total sequence of n repetitions, let us select the sub-sequence which consists of the f_A repetitions where the event A has occurred. In this sub-sequence there are exactly f_{AB} repetitions where the event B has occurred, since these are precisely the f_{AB} repetitions in the original sequence where both A and B have occurred. The ratio f_{AB}/f_A will thus be the frequency ratio of the event B *within the selected sub-sequence*, which we may conveniently denote as the *conditional frequency ratio of B, relative to the hypothesis that A has occurred*. — In a similar way f_{AB}/f_B will be the conditional frequency ratio of A, relative to the hypothesis that B has occurred.

Ex. 1. — Suppose that we have observed a group of $n = 100$ newborn children with respect to the occurrence or non-occurrence of the events $A = $ "boy" and $B = $ "blue eyes". If we have found $f_A = 51$ boys, $f_B = 68$ blue-eyed children and $f_{AB} = 34$ blue-eyed boys, the results of our observations can be arranged in the following table:

	Blue	Not blue	Total
Boys	34	17	51
Girls	34	15	49
Total	68	32	100

In the total sequence of 100 observations we thus find that the frequency ratio for boys is $f_A/n = 0.51$, while the frequency ratio for blue-eyed children is $f_B/n = 0.68$. Further, the conditional frequency ratio of boys among the blue-eyed, i.e., the frequency ratio for boys within the sub-sequence formed by the $f_B = 68$ blue-eyed children, is $f_{AB}/f_B = \frac{34}{68} = 0.50$, while the conditional frequency ratio for the blue-eyed among the boys is

$$f_{AB}/f_B = \frac{34}{51} = 0.67.$$

We now have the following identical relation:

$$\frac{f_{AB}}{f_A} = \frac{f_{AB}}{n} : \frac{f_A}{n},$$

and for large values of n it can be expected that the frequency ratios f_{AB}/n and f_A/n will be approximately equal to the corresponding probabilities $P(AB)$ and $P(A)$ respectively. Now, if we suppose that $P(A)$ is different from zero, and if we introduce the new symbol

$$(3.4.1) \qquad P(B \mid A) = \frac{P(AB)}{P(A)},$$

it follows that we should expect that, for large values of n, the conditional frequency ratio f_{AB}/f_A will be approximately equal to the quantity $P(B \mid A)$. This quantity thus constitutes an ideal value for the conditional frequency ratio, in the same way as, e.g., the probability $P(A)$ constitutes an ideal value for the frequency ratio f_A/n.

Accordingly the quantity $P(B \mid A)$ defined by 3.4.1 will be called the *conditional probability of B, relative to the hypothesis that A has occurred.* — In a similar way, when $P(B)$ is different from zero, the quantity

$$(3.4.2) \qquad P(A \mid B) = \frac{P(AB)}{P(B)}$$

will be the *conditional probability of A, relative to the hypothesis that B has occurred.*[1]

From 3.4.1 and 3.4.2 we obtain the following formula, which is commonly known as the *multiplication rule* of probability theory:

$$(3.4.3) \qquad P(AB) = P(A) P(B \mid A)$$
$$= P(B) P(A \mid B).$$

Ex. 2. — The conditional probabilities can often be determined directly, by means of the symmetry principle on which the classical probability definition is founded. This can be done, e.g., in the following simple example. Two cards C_1 and C_2 are drawn simultaneously from the same set. A denotes the event that C_1 is a heart, B the event that C_2 is a heart. Obviously we then have $P(A) = \dfrac{13}{52}$. Further, if A has occurred, C_2 will be a card drawn at random from a set containing 51 cards, 12 of which are hearts,

---

[1] It should be observed that the properties of frequency ratios are not here used in order to deduce some new basic property of the probabilities, but only to *justify the terminology* that will be used in connection with the new symbols $P(B \mid A)$ and $P(A \mid B)$.

so that $P(B \mid A) = \dfrac{12}{51}$. The probability that both cards are hearts is then, according to the multiplication rule, $P(AB) = \dfrac{13}{52} \cdot \dfrac{12}{51} = \dfrac{1}{17}$.

The multiplication rule is directly extended to the case of more than two events, in the same way as this was done in 3.2 for the addition rule. The general multiplication rule for an arbitrary number of events can be written in the following form, where the notations should be easily understood:

(3.4.4) $P(A_1 A_2 \ldots A_m) =$

$$= P(A_1) P(A_2 \mid A_1) P(A_3 \mid A_1 A_2) \ldots P(A_m \mid A_1 A_2 \ldots A_{m-1}).$$

3.5. Independent events. — Let the events A and B have probabilities $P(A)$ and $P(B)$, both of which are different from zero. As soon as one of the two relations

$$P(A \mid B) = P(A), \quad \text{or} \quad P(B \mid A) = P(B)$$

holds, it follows from 3.4.3 that the other one holds as well, and that we have in addition

(3.5.1) $$P(AB) = P(A) P(B).$$

In this case the probability of any of the events A and B is independent of whether the other event has occurred or not, and we shall accordingly say that A and B are *independent events* from the point of view of probability theory. Sometimes this will be expressed by saying that A and B are *stochastically independent* events.[1] Thus for independent events the multiplication rule takes a particularly simple form, just as did the addition rule for mutually exclusive events.

If from our knowledge of the nature of the phenomenon under investigation we judge two events A and B to be *causally independent*, i.e., if we are not able to trace any causal connection between the events, it seems natural to assume that the relations given above will hold, and that the events will thus also be independent from the point of view of probability theory, or stochastically independent, in the sense given above.

---

[1] The words "stochastic" and "stochastical", which are derived from a Greek word, are now fairly commonly used in the literature, meaning simply "connected with random experiments and probability".

Ex. — Let us modify the example given in connection with 3.4.3 so that the cards C_1 and C_2 are drawn from two different sets of cards. If both drawings are performed under the same a priori given conditions, we should feel inclined to assume that the two events A and B will be causally independent, and thus also stochastically independent. By 3.5.1 the probability that both cards are hearts will then in this case be

$$P(A\,B) = \frac{13}{52} \cdot \frac{13}{52} = \frac{1}{16}.$$

Generally, let m events $A_1, \ldots, A_m$ be given, and suppose that $P(A_\nu) \neq 0$ for all $\nu = 1, 2, \ldots, m$, and that for any group $A_{\nu_1}, \ldots, A_{\nu_k}$ of $k \leq m$ among the events A_ν we have

(3.5.2) $$P(A_{\nu_1} \mid A_{\nu_2} \ldots A_{\nu_k}) = P(A_{\nu_1}).$$

The events $A_1, \ldots, A_m$ will then be called a group of m *stochastically independent* events, since the probability of any of these events does not depend on whether some of the other events have occurred or not. In this case, it is readily seen that the multiplication rule 3.4.4 will assume the simpler form

(3.5.3) $$P(A_1 A_2 \ldots A_m) = P(A_1)\,P(A_2) \ldots P(A_m),$$

which is a generalization of 3.5.1.

It should be well observed that the simpler multiplication formulae 3.5.1 and 3.5.3 have been proved only for the case of independent events. As long as we deal with problems concerning games of chance it will in general not be difficult to decide whether this condition is satisfied or not. In other cases, however, this may sometimes be very difficult, and thus it will always be necessary to use great caution in the application of these formulae. As soon as it cannot be made perfectly clear that we are concerned with independent events, the general formulae 3.4.3 and 3.4.4 should be applied. — Similar remarks hold in respect of the simplified addition rules 3.2.1 a and 3.2.3 a, which have been proved only for the case of mutually exclusive events.

3.6. Bayes' theorem. — It follows from 3.4.3 that, if $P(A)$ and $P(B)$ are both different from zero, we shall have

(3.6.1) $$P(A \mid B) = \frac{P(A)\,P(B \mid A)}{P(B)}.$$

Suppose now that B can only occur in combination with one of m events $A_1, \ldots, A_m$, any two of which are mutually exclusive, and let us write

$$P(A_\nu) = \pi_\nu, \quad P(B \mid A_\nu) = p_\nu, \quad (\nu = 1, 2, \ldots, m).$$

We then have

$$B = A_1 B + A_2 B + \cdots + A_m B,$$

where any two of the events in the second member are mutually exclusive. By 3.2.3 a and 3.4.3 we then obtain

$$P(B) = \pi_1 p_1 + \pi_2 p_2 + \cdots + \pi_m p_m.$$

If now we replace in 3.6.1 the event A by an arbitrary A_ν, we shall have

$$(3.6.2) \qquad P(A_\nu \mid B) = \frac{\pi_\nu p_\nu}{\pi_1 p_1 + \cdots + \pi_m p_m}.$$

This formula is known under the name of BAYES' *theorem*.

This theorem is often applied in the following general situation. At an actual performance of a certain random experiment we have observed the occurrence of the event B, which may occur only as an effect of one of m mutually exclusive causes $A_1, \ldots, A_m$. The probability π_ν that the cause A_ν is acting, calculated before the performance of the experiment, will be denoted as the *a priori probability* of A_ν. Further, under the hypothesis that the cause A_ν is acting, we have a certain probability p_ν that this cause will lead to the occurrence of B. The formula 3.6.2 then gives an expression of the *a posteriori probability* of A_ν, i.e., the probability $P(A_\nu \mid B)$ that the cause A_ν is acting, *calculated under the hypothesis that we have observed the occurrence of B.*

The main difficulty encountered in the majority of statistical applications of BAYES' theorem arises from the fact that in most cases the a priori probabilities π_ν are entirely unknown. This is a serious difficulty which considerably reduces the practical importance of the theorem. We shall later see examples of cases where this difficulty arises (16.1). In this section we shall only give a simple example of a perfectly legitimate application of the theorem.

Ex. — In a factory two machines M_1 and M_2 are used for the manufacture of screws. We suppose that each screw can be uniquely classified as "good" or "bad". M_1 produces per day n_1 boxes of screws of which on the average p_1 % are bad, while the corresponding numbers for M_2 are n_2 and p_2. From the total production of both machines for a certain day we choose at random a box, and from this box we draw a screw which on investigation proves to be bad. What is the probability that we have chosen a box manufactured by M_1?

If we suppose that all boxes are similar with respect to shape, colour, etc., we can regard the $n_1 + n_2$ boxes from both machines as symmetrical cases in the sense of the classical probability definition. The a priori probability that a randomly chosen box comes from M_1 will then be $\dfrac{n_1}{n_1 + n_2}$, and similarly for M_2. On the other hand, the probability to draw a bad screw from a box made by M_1 is $p_1/100$, while the corresponding probability for a box from M_2 is $p_2/100$. The a posteriori probability that the selected box comes from M_1, *when we have observed that the drawn screw is bad*, will then according to 3.6.2 be equal to $\dfrac{n_1 p_1}{n_1 p_1 + n_2 p_2}$.

Suppose, e.g., that $p_1 > p_2$. Then

$$\frac{n_1 p_1}{n_1 p_1 + n_2 p_2} - \frac{n_1}{n_1 + n_2} = \frac{n_1 n_2 (p_1 - p_2)}{(n_1 + n_2)(n_1 p_1 + n_2 p_2)} > 0.$$

As might have been expected, the observation that we have drawn a bad screw will thus in this case tend to increase the probability that the selected box comes from M_1.

3.7. Problems. — 1.
Ten persons are arranged at random a) in a row, b) in a ring. All possible arrangements are assumed to have the same probability. Find the probability that two given persons will be next to one another.

2. On an empty chess-board the white and the black queen are placed at random. All possible positions are assumed to have the same probability. Find the probability that the two queens will be in striking position.

3. Two cards are drawn simultaneously from the same set of 52 cards. Find the probability that at least one of them will be a heart. (Use 3.2.1 and 3.4, ex. 2.)

4. U_1 and U_2 are two similar urns. U_1 contains n_1 balls, ν_1 of which are white, while U_2 contains n_2 balls, ν_2 of which are white. A person chooses at random an urn and draws a ball from it. Find the probability that he will draw a white ball. (We assume that the probability of choosing U_1 is $\frac{1}{2}$, and the same for U_2. Begin by finding the probability of the event that U_1 will be selected *and* a white ball drawn from it, and then find the same probability for U_2. Criticize the following argument: there are $n_1 + n_2$ possible cases, represented by the $n_1 + n_2$ balls in the two urns; of these $\nu_1 + \nu_2$ are favourable to the event of drawing a white ball, hence the required probability is $(\nu_1 + \nu_2)/(n_1 + n_2)$.

5. A and B are two events connected with a random experiment E. The probabilities $P(A)$, $P(B)$, and $P(AB)$ are given. Find expressions in terms of these probabilities for the probabilities of the following events: a) $A^* + B^*$, b) $A^* B^*$, c) $A^* + B$, d) $A^* B$, e) $(A + B)^*$, f) $(AB)^*$, g) $A^*(A + B)$, h) $A + A^* B$. (Hint: In a) and b), begin by finding the opposite event and its probability.)

6. Let p_ν denote the probability that exactly ν of the events A and B will occur $(\nu = 0, 1, 2)$. Find expressions of p_0, p_1 and p_2 in terms of $P(A)$, $P(B)$ and $P(AB)$.

7. Generalize the preceding exercise to the case of three events A, B, and C.

8. One shot is fired from each of three guns, the probability of a hit being 0.1, 0.2 and 0.3 respectively. Find the probability of each of the possible number of hits.

9. Prove that for any two events A and B we have $P(AB) \leqq P(A) \leqq P(A + B) \leqq P(A) + P(B)$.

10. A and B are independent events, both of which are neither certain nor impossible. Prove that the opposite events A^* and B^* are independent.

11. $A_1, \ldots, A_n$ are independent events. Show that

$$P(A_1 + \cdots + A_n) = 1 - P(A_1^*) P(A_2^*) \ldots P(A_n^*).$$

12. A, B, and C are three events, neither of which is impossible. Show that, in order that the independence conditions 3.5.2 should be satisfied, it is sufficient to require that $P(AB) = P(A) P(B)$, $P(AC) = P(A) P(C)$, $P(BC) = P(B) P(C)$, and $P(ABC) = P(A) P(B) P(C)$.

13. It might be thought that, in the preceding exercise, it would be sufficient to assume the truth of the first three relations. Three events would then be independent if, and only if, they are pairwise independent. However, this conjecture is not true, as the following example shows. — Each of the numbers 1, 2, 3, and 123 is inscribed on one of four tickets, one of which is drawn at random. Let A, B, and C denote the events that the number drawn will contain a one, a two, or a three respectively. Find the probability of the simultaneous occurrence of any group of one, two, or three among the events A, B, and C. Show that a) any pair of the events are independent, while b) the three events A, B, C are *not* independent.

14. An urn contains two balls. It is known that the urn was filled by tossing a coin twice and putting a white ball in the urn for each head and a black ball for each tail. A ball is drawn from the urn and is found to be white. Find the probability that the other ball in the urn is also white.

15. *Applications to mortality tables.* Let the probability that a person aged x years will survive one year be denoted by p_x, while $q_x = 1 - p_x$ is the probability that he will die within a year. Further, let $_t p_x$ denote the probability that he will survive t years. Show that

$$_{t+u} p_x = {}_t p_x \cdot {}_u p_{x+t},$$

and that, when t is an integer n, we have

$$_n p_x = p_x \, p_{x+1} \cdots p_{x+n-1}.$$

Let l_0 be an arbitrary positive constant, and define the functions l_x and d_x by writing for all $x > 0$

$$l_x = l_0 \cdot {}_x p_0, \quad d_x = l_x - l_{x+1}.$$

Show that

$$_t p_x = \frac{l_{x+t}}{l_x}, \quad q_x = \frac{d_x}{l_x}.$$

From a single table giving the function l_x for $x = 0, 1, \ldots$, we can thus simply compute all probabilities q_x and $_n p_x$ for integral values of x and n. The probability q_x is known as the *rate of mortality* at age x.

CHAPTER 4

SOME APPLICATIONS OF THE FUNDAMENTAL THEOREMS

4.1. Repeated observations. — Consider a sequence of n repetitions of a given random experiment E, and let us each time observe whether a given event A takes place or not. We assume that all n observations are performed under uniform conditions, meaning that all conditions which, by previous experience or a priori considerations, may be expected to be relevant to the outcome of the experiment, should be kept constant during the whole sequence of observations.

The probability of the occurrence of A must then be supposed to have the same value $P(A) = p$ for all n observations. Thus if A_ν denotes the event: "A occurs in the ν:th observation", we shall have $P(A_\nu) = P(A) = p$ for every ν. Further, the value of this probability will remain unaffected by any information as to whether A has occurred or failed to occur in any group of the other observations. By 3.5.2 the events $A_1, \ldots, A_n$ will then be independent events, and according to 3.5.3 the probability that A occurs *in all n observations* will be equal to p^n.

On the other hand, if we denote the probability that A *fails to occur* in one of our observations by $P(A^*) = q$, we have according to 3.2.4 $p + q = 1$, or $q = 1 - p$. In the same way as above it is then seen that the probability that A fails to occur *in all n observations* will be equal to $q^n = (1 - p)^n$. The opposite event, which implies that A occurs *in at least one of the n observations*, will then by 3.2.4 have the probability $1 - q^n$.

Ex. 1. — By the aid of the last formula we can discuss one of DE MÉRÉ's questions to PASCAL (1.1). From his practical experience DE MÉRÉ had found that it was favourable to make a one-to-one bet to have at least one six in four throws with an ordinary die, while it was unfavourable to bet on the same conditions to have at least one double-six in 24 throws with two dice. In a long series of bets of the former kind, the frequency ratio of his gains had in fact proved to be larger than $\frac{1}{2}$, i.e., the number of gains exceeded the number of losses, so that the total result was a gain for the gambler. On the other hand, in bets of the latter kind the opposite circumstances appeared, so that the total result was a loss.

DE MÉRÉ regarded these results as contradictory to mathematical calculation. His argument was that, since 4 is to 6 (the total number of possible cases at a throw with one die) as 24 to 36 (the total number of possible cases at a throw with two dice) the chances of a gain should be equal in both cases.

However, according to our formula the probability of having at least one six in four throws with one die is

$$1 - \left(\frac{5}{6}\right)^4 = 0.5177,$$

while the probability of having at least one double-six in 24 throws with two dice is

$$1 - \left(\frac{35}{36}\right)^{24} = 0.4914.$$

As the first probability is larger than $\frac{1}{2}$, while the second is smaller than $\frac{1}{2}$, the first bet is favourable while the second is unfavourable, which entirely agrees with DE MÉRÉ's experience. If we modify the conditions of the second bet by making 25 throws instead of 24, the probability of a gain will be 0.5055, and the bet will be favourable.

We now proceed to deduce a very important formula, which will be repeatedly used in the sequel, and which contains the formulae given above as simple particular cases.

In the sequence of n observations considered above, it is obvious that the probability that A occurs in a certain observation, as well as the probability that A fails to occur in the same observation, is entirely independent of the number of times that A may have occurred or failed to occur in the other observations. We shall accordingly say that we are concerned with a sequence of n *independent observations*.

We have defined the frequency f of A in our sequence of n observations as the number of occurrences of A during the whole sequence. The frequency f can assume any of the values $0, 1, \ldots, n$. Let now ν be an arbitrary value chosen among these. What is the probability that the frequency f will assume exactly the given value ν, i.e., the probability that the event A occurs exactly ν times in the course of our n independent observations?

If we register the results of our n observations by writing A for each observation where A has occurred, and otherwise A^*, we shall have a list of, e.g., the following appearance:

$$A^* A^* A A^* A A A \ldots A^* A^* A.$$

Suppose that the number of A:s contained in the list written here as an example is equal to ν. The number of A^*:s will then obviously be $n - \nu$. What is the probability that the registered results of our sequence will be identical with this particular list?

As we are dealing with independent events, we may use the multiplica-

43

tion rule in the simplified form 3.5.3, and thus find the following expression for the required probability

$$q\,q\,p\,q\,p\,p\,p \ldots q\,q\,p = p^{\nu}\,n^{n-\nu}.$$

However, what we really want to calculate is not the probability that the results of our sequence will agree with the particular list written above, but the probability that these results will be expressed by *any* list containing A in exactly ν places and A^* in the remaining $n-\nu$ places. As obviously two different lists always represent mutually exclusive events, it follows from the simplified addition rule 3.2.3a that this probability is equal to the sum of the probabilities of all sequences corresponding to arrangements of the desired type. For each particular list we have just found that the probability is $p^{\nu}q^{n-\nu}$, and thus it will only be necessary to find the number of possible different arrangements containing A in ν places and A^* in $n-\nu$ places. In other words: in how many different ways is it possible to choose from among the total number of n places in our list the ν places where we write A? — It is well known from the elements of combinatorial analysis that this choice can be made in

$$\binom{n}{\nu} = \frac{n!}{\nu!\,(n-\nu)!} = \frac{n\,(n-1)\,\ldots\,(n-\nu+1)}{1\cdot2\cdot\ldots\cdot\nu}$$

different ways, and the answer to our problem will thus be that the required probability that A will occur exactly ν times is

(4.1.1)
$$\binom{n}{\nu}\,p^{\nu}\,q^{n-\nu}.$$

Let us now try to calculate the probability that the number of occurrences of A will lie *between two given limits* ν_1 and ν_2, the limits included, i.e., the probability that the frequency f of A satisfies the inequality $\nu_1 \leq f \leq \nu_2$. As two different values of ν will always represent mutually exclusive events, we may again calculate the required probability by addition and thus find the expression

(4.1.2)
$$\sum_{\nu=\nu_1}^{\nu_2}\binom{n}{\nu}\,p^{\nu}\,q^{n-\nu}.$$

Taking here in particular $\nu_1 = 0$ and $\nu_2 = n$, we shall have the probability of a *certain* event, since the frequency f will always be sure to assume one of the values $0, 1, \ldots, n$. In fact, we have according to the binomial theorem

(4.1.3)
$$\sum_{\nu=0}^{n}\binom{n}{\nu}\,p^{\nu}\,q^{n-\nu} = (p+q)^{n} = 1,$$

which agrees with the general result in 3.1 that the probability of a certain event is always equal to unity. On account of this connection with the binomial theorem, the important probability formula 4.1.1 is known as the *binomial formula*.

Ex. 2. — The probability π_ν of having heads exactly ν times in the course of $2\,n$ throws with a coin is according to 4.1.1

$$\pi_\nu = \binom{2\,n}{\nu}\left(\frac{1}{2}\right)^\nu\left(\frac{1}{2}\right)^{2\,n-\nu} = \binom{2\,n}{\nu}\left(\frac{1}{2}\right)^{2\,n}.$$

By means of the well-known properties of the binomial coefficients, it is easily seen that for a given value of n this probability attains its maximum when $\nu = n$, and that this maximum probability, i.e., the probability π_n of obtaining exactly n heads and n tails in the $2\,n$ throws, is

$$\pi_n = \binom{2\,n}{n}\left(\frac{1}{2}\right)^{2\,n} = \frac{(2\,n)!}{2^{2\,n}\,(n!)^2} = \left(1-\frac{1}{2}\right)\left(1-\frac{1}{4}\right)\cdots\left(1-\frac{1}{2\,n}\right).$$

However, for $x > 0$ we have $1 - x < e^{-x}$, and thus

$$\pi_n < e^{-\frac{1}{2}\left(1+\frac{1}{2}+\cdots+\frac{1}{n}\right)} < e^{-\frac{1}{2}\log n} = \frac{1}{\sqrt{n}}.$$

It follows that, when n tends to infinity, the maximum probability π_n, and a fortiori every other π_ν, will tend to zero. We shall later (cf. 6.4) deduce more accurate and more general results concerning the behaviour of the binomial formula for large values of n.

Ex. 3. — The reader should use 4.1.2 to deduce the expression $1 - q^n$ given above for the probability that the event A will occur at least once in the course of a sequence of n observations.

4.2. Drawings with and without replacement. — In a great number of important applications of probability theory the following problem is encountered. We have a group of $N = N_1 + N_2$ objects of two different kinds, N_1 of the "first kind" and N_2 of the "second kind". Without being able to distinguish between objects of the two different kinds, we draw a sample of n objects from the total group of N objects, regarding the latter as symmetrical cases in the sense of the classical definition. What is the probability of obtaining a sample consisting of ν objects of the first kind and $n - \nu$ of the second kind?

Problems of this type arise, e.g., in connection with the drawing of cards from a set, or of balls from an urn. They also arise in connection with practically important applications, such as the selection of a sample from a batch of manufactured articles, a number of families for a representative social investigation, or a number of subjects to be interviewed for a public opinion poll, etc.

Although the following discussion will be perfectly general, it will be convenient to use one particular example to illustrate the argument, and we shall choose the example of drawing balls from an urn. Thus we suppose that we are given an urn containing $N = N_1 + N_2$ balls of equal size and shape, of which N_1 are white and N_2 black. Without being able to distinguish between balls of different colours, we draw n balls from the urn, and require the probability of having ν white and $n - \nu$ black balls.

The problem as stated in this form is not yet sufficiently precise to admit a unique solution. In the applications the problem usually presents itself in two different forms, which in terms of our illustrative example may be described as drawings *with replacement* or *without replacement*. In the former case the n balls are drawn one by one and every drawn ball is replaced in the urn as soon as its colour has been noted, and before the next drawing is made. In the latter case, on the other hand, no such replacement is made, which amounts to saying that all the n balls are simultaneously drawn from the urn.

Drawings with replacement. — In this case the n drawings may be regarded as a sequence of n independent observations performed under uniform conditions. Thus we are concerned with a particular case of the problem treated in the preceding section. At each drawing the probability of obtaining a white ball is $p = N_1/N$, and the probability of having ν white and $n - \nu$ black balls is according to the binomial formula

$$\binom{n}{\nu} p^\nu q^{n-\nu} = \binom{n}{\nu} \left(\frac{N_1}{N}\right)^\nu \left(\frac{N_2}{N}\right)^{n-\nu}.$$

Drawings without replacement. — This is the case usually encountered in practical applications, when a sample of n is drawn from a group or *population* consisting of N objects or individuals. Accordingly this case is particularly important from the point of view of the theory and technique of *statistical sampling,* and we shall later (16.5) return to a more detailed discussion of questions belonging to this order of ideas.

In this case we must obviously have $n \leq N$. Further, it is clear that the required probability will be equal to zero as soon as either $\nu > N_1$ or $n - \nu > N_2$, since the number of white or black balls in our sample can never exceed the corresponding number in the total population consisting of all balls in the urn.

For all other values of ν, we shall now find the required probability by means of a direct application of the classical probability definition. The possible cases are all groups of n which can be formed by means of the N balls in the urn. The number of these groups is $\binom{N}{n}$. The favourable cases are formed by selecting in all possible ways a group of ν among the N_1 white balls, and combining each such group with all possible groups of $n-\nu$ selected among the N_2 black balls. The required probability of having ν white and $n-\nu$ black balls will thus be

(4.2.1)
$$\frac{\binom{N_1}{\nu}\binom{N_2}{n-\nu}}{\binom{N}{n}}.$$

After some algebraical reductions this expression can also be written in the following form

$$\binom{n}{\nu}\left(\frac{N_1}{N}\right)^{\nu}\left(\frac{N_2}{N}\right)^{n-\nu}.$$

$$\cdot\frac{\left(1-\frac{1}{N_1}\right)\left(1-\frac{2}{N_1}\right)\cdots\left(1-\frac{\nu-1}{N_1}\right)\left(1-\frac{1}{N_2}\right)\left(1-\frac{2}{N_2}\right)\cdots\left(1-\frac{n-\nu-1}{N_2}\right)}{\left(1-\frac{1}{N}\right)\left(1-\frac{2}{N}\right)\cdots\left(1-\frac{n-1}{N}\right)}.$$

If in this expression we allow N, N_1 and N_2 to tend to infinity in such a way that N_1/N tends to a limiting value p, while n and ν are kept fixed, we find without difficulty that the expression will tend to the limiting value $\binom{n}{\nu}p^{\nu}q^{n-\nu}$, which is the same formula as the one obtained for drawings with replacement with the probability p. A sequence of n drawings with replacement can thus be regarded as a sample of n individuals, drawn without replacement *from an infinite population*.

Ex. 1. — The probability π_{ν} that, at a deal for bridge, a given player will obtain exactly ν aces is by 4.2.1

$$\pi_{\nu}=\frac{\binom{4}{\nu}\binom{48}{13-\nu}}{\binom{52}{13}}.$$

From this formula we obtain the following values

47

$$\pi_0 = \frac{6\,327}{20\,825} = 0.304, \quad \pi_1 = \frac{9\,139}{20\,825} = 0.439,$$

$$\pi_2 = \frac{4\,446}{20\,825} = 0.213, \quad \pi_3 = \frac{858}{20\,825} = 0.041, \quad \pi_4 = \frac{55}{20\,825} = 0.003.$$

As a check on the computation we find that the sum of all five probabilities is equal to 1.

Ex. 2. — A manufactured article is delivered by the producer in batches of N apparently equal units, each of which may on a detailed inspection be classified as "good" or "bad" according to certain fixed specifications. A certain consumer who is anxious to admit only batches containing less than N_0 bad units, uses the following procedure for testing the quality of the batches as they are delivered. From each batch a sample of n units is drawn at random and inspected. If all n sample units are good, the whole batch is admitted without further inspection, while in the opposite case the batch is rejected. What can be said about the risk involved in this procedure of admitting a "bad" batch, i.e., a batch with at least N_0 bad units?

If the unknown true number of bad units in a certain batch is denoted by N_1, the probability that all n units in the sample will be good is according to 4.2.1

$$\frac{\binom{N-N_1}{n}}{\binom{N}{n}} = \left(1 - \frac{N_1}{N}\right)\left(1 - \frac{N_1}{N-1}\right) \cdots \left(1 - \frac{N_1}{N-n+1}\right).$$

If the batch is "bad" from the point of view of the consumer, i.e., if $N_1 \geq N_0$, this probability is obviously at most equal to

$$\left(1 - \frac{N_0}{N}\right)\left(1 - \frac{N_0}{N-1}\right) \cdots \left(1 - \frac{N_0}{N-n+1}\right).$$

This expression thus represents the maximum risk of admitting a bad batch which is attached to the testing method used by the consumer. Choosing, e.g., $N = 100$, $N_0 = 10$, we obtain the following values:

n	Maximum risk, percent.
10	33.0
20	9.5
30	2.3

Quality control by means of sampling inspection has very important applications in industry, though of course the methods used in practice will be more complicated than in the simple example discussed here. We shall return to the subject in 16.6.

Drawings with more than two alternatives. — The preceding problem can be generalized so that we consider an urn containing balls of more than two different colours. Let the urn contain $N = N_1 + \cdots + N_k$ balls of k different colours, where N_r denotes the number of balls of the r:th colour. As before we draw n balls from the urn and we require the probability of having ν_1 balls of the first colour, ν_2 balls of the second colour, etc., where $\nu_1 + \cdots + \nu_k = n$.

In the case of drawings with replacement a similar argument as in the case of the binomial formula will lead to the expression

$$(4.2.2) \qquad \frac{n!}{\nu_1! \cdots \nu_k!} \, p_1^{\nu_1} \cdots p_k^{\nu_k}$$

for the required probability. Here $p_r = N_r/N$ denotes the probability of obtaining in an arbitrary drawing a ball of the colour having the number r. If we sum this expression over all non-negative integral values of $\nu_1, \ldots, \nu_k$ having the sum n, we obtain the identity

$$\sum_{\nu_1 + \cdots + \nu_k = n} \frac{n!}{\nu_1! \cdots \nu_k!} \, p_1^{\nu_1} \cdots p_k^{\nu_k} = (p_1 + \cdots + p_k)^n = 1,$$

which is a generalization of 4.1.3. Accordingly the formula 4.2.2 is known as the *polynomial formula*.

In the case of drawings without replacement, a straightforward generalization of the above deduction of 4.2.1 will give the expression

$$(4.2.3) \qquad \frac{\binom{N_1}{\nu_1} \binom{N_2}{\nu_2} \cdots \binom{N_k}{\nu_k}}{\binom{N}{n}}$$

for the required probability.

4.3. Some further problems on games of chance. — 1. *The problem of points.* — This is one of the problems originally raised by the chevalier DE MÉRÉ and discussed in the correspondence between PASCAL and FERMAT (cf. 1.1). In TODHUNTER's History (se list of references) the problem is stated in the following terms. Two players want each a given number of points in order to win; if they separate without playing out the game, how should the stakes be divided between them?

Suppose that the players A and B are engaged in a sequence of repetitions of a given game, where in each repetition the probability of A's winning is p, while the corresponding probability for B is $q = 1 - p$. Each gain counts as one point, and the game is interrupted at a moment when the number of points still required in order to win the total stakes is m for A, and n for B.

It seems reasonable that under these circumstances the stakes should be divided in proportion to the respective probabilities of winning, if the game were going to be continued. In order to find these probabilities we suppose that the game will be repeated an additional number of $m+n-1$ times. This must either give at least m points to A, or at least n points to B. Further, if A obtains at least m points in the course of these $m+n-1$ games, B will obtain less than n points, and vice versa. It follows that the event "A wins the stakes" will be equivalent to the event "A wins at least m of the $m+n-1$ games", and similarly for B. From 4.1.2 we then directly obtain the following expression for the required probability that A wins the stakes:

$$p_A = \sum_{\nu=m}^{m+n-1} \binom{m+n-1}{\nu} p^\nu q^{m+n-1-\nu}.$$

The corresponding probability for B is obviously $p_B = 1 - p_A$, and it follows that the stakes should be divided in the proportion $p_A : p_B$.

2. A problem in bridge. — Find the probability that at a deal for bridge at least one of the four players will have thirteen cards of the same suit.

Let the four players be A, B, C and D, and let A also stand for the event, "the player A obtains thirteen cards of the same suit", and similarly in respect of B, C and D. We then have to find the probability that a least one of the events A, B, C and D will occur. According to the notation introduced in 3.2, this probability should be denoted by $P(A+B+C+D)$.

This probability can be expressed by means of the formula 3.2.3. On account of the symmetry we shall then have

$$P(A) = P(B) = P(C) = P(D),$$

$$P(AB) = P(AC) = P(AD) = P(BC) = P(BD) = P(CD),$$

$$P(ABC) = P(ABD) = P(ACD) = P(BCD),$$

and thus 3.2.3 gives

$$P(A+B+C+D) = 4P(A) - 6P(AB) + 4P(ABC) - P(ABCD).$$

However, it is readily seen that the events ABC and $ABCD$ are equivalent. In fact, if the players A, B and C all have thirteen cards of the same suit, the same thing must hold for D. Thus we have

$$P(A+B+C+D) = 4P(A) - 6P(AB) + 3P(ABC).$$

The thirteen cards dealt to the player A can be selected from the set in $\binom{52}{13}$ different ways, four of which are favourable to the event A, since this event will occur if the player obtains all thirteen cards of any of the four suits in the set. Hence

$$P(A) = \frac{4}{\binom{52}{13}}.$$

If A has obtained a complete suit, the 39 cards of the three remaining suits are still left to be disposed, and among these the thirteen cards dealt to B can be selected in $\binom{39}{13}$ different ways, three of which are favourable to the event B. The multiplication rule 3.4.3 then gives

$$P(AB) = \frac{4}{\binom{52}{13}} \cdot \frac{3}{\binom{39}{13}}.$$

Using the multiplication rule in the general form 3.4.4 we find in the same way

$$P(ABC) = \frac{4}{\binom{52}{13}} \cdot \frac{3}{\binom{39}{13}} \cdot \frac{2}{\binom{26}{13}}.$$

The final expression of the required probability thus becomes

$$P(A+B+C+D) = \frac{16\binom{39}{13}\binom{26}{13} - 72\binom{26}{13} + 72}{\binom{52}{13}\binom{39}{13}\binom{26}{13}},$$

which is approximately equal to $0.25 \cdot 10^{-10}$. This is a good instance of an event of the type which, in 3.1, we denoted as an *unusual event*. If in a bridge club there are 100 deals made each day, the event here considered will, on the average, occur a little less often than once in every million years, always supposing that the cards are properly mixed before each deal! If we are invited to the club for, say, one day, it seems to be a practically safe conclusion that we shall not have the occasion of seeing this event occur.

3. *The game of rencontre.* — An urn contains n tickets numbered 1, 2, ..., n. The tickets are drawn one by one without replacement, and if the ticket numbered r appears in the r:th drawing, this is denoted as a *match* or a *rencontre*. Find the probability of having at least one rencontre!

This is one of the classical problems of the theory of games, and was treated for the first time by MONTMORT in 1708. — Let A_r denote the event that there will be a rencontre in the r:th drawing. The required probability of at least one rencontre is then $P(A_1 + \cdots + A_n)$ and we express this by means of the addition formula 3.2.3. We shall first calculate an arbitrary term $P(A_{r_1} A_{r_2} \ldots A_{r_k})$ in this formula, i.e., the probability that the drawings numbered $r_1, \ldots, r_k$ will all give rencontres. The $n!$ possible arrangements of the n tickets give us $n!$ possible cases. The favourable cases are the arrangements where the numbers $r_1, \ldots, r_k$ appear in their right places. The remaining $n - k$ numbers can then occur in $(n - k)!$ different arrangements, so that the number of favourable cases is $(n-k)!$, and we have

$$P(A_{r_1} A_{r_2} \ldots A_{r_k}) = \frac{(n-k)!}{n!}.$$

The number of those terms in the second member of 3.2.3 which contain products of k events is obviously $\binom{n}{k}$, and thus we obtain the following expression of the required probability:

$$P(A_1 + \cdots + A_n) = \binom{n}{1}\frac{(n-1)!}{n!} - \binom{n}{2}\frac{(n-2)!}{n!} + \cdots + (-1)^{n-1}\binom{n}{n}\frac{1}{n!}$$

$$= \frac{1}{1!} - \frac{1}{2!} + \frac{1}{3!} - \cdots + (-1)^{n-1}\frac{1}{n!}.$$

When n tends to infinity, this probability tends to the limiting value $1 - e^{-1} = 0.63212 \ldots$. The approximation to this limit is very good even for moderate values of n, thus for $n = 7$ the probability of at least one rencontre is $0.63214 \ldots$.

4.4. Problems. — 1. Find the probability of having exactly three heads in five throws with a coin.

2. Assuming that each child has the probability 0.51 of being a boy, find the probability that a family with six children should have a) at least one boy, b) at least one girl.

3. Find the probability of obtaining exactly one heart in four drawings from an ordinary set of cards, when the drawings are made a) with replacement, b) without replacement.

4. Find the probability of hitting a ship when one disposes of four torpedos, each with the probability $\frac{1}{4}$ of making a hit.

5. An ordinary die is thrown five times. Find the most probable number of sixes.

6. Two persons play heads and tails and have agreed to continue the game until heads and tails have both occurred at least three times. Find the probability that the game will not be finished when ten throws have been made.

7. r objects are distributed at random in n cells, all possible arrangements having the same probability. Find the probability that a given cell contains exactly k objects.

8. Each coefficient of the equation $ax^2 + bx + c = 0$ is determined by means of a throw with an ordinary die. Find the probability that the equation will have a) real roots, b) rational roots.

9. Six throws are made with a coin. Find the probability that there will be at least one uninterrupted run of three heads.

10. In a lottery of 400 tickets there are four prizes. A person buys ten tickets. Find his probability of obtaining at least one prize.

11. In a lottery of 10 000 tickets there are 100 prizes. A person desires to buy so many tickets that his probability of obtaining at least one prize will exceed 50 %. How many tickets will it be necessary for him to buy?

12. At a deal for bridge the player A has received two aces. Find the probability for each of the possible numbers of aces that may have been dealt to his partner.

13. At a deal for bridge the total number of trump cards held by the two partners A and B is nine. Find the probabilities that the remaining four trump cards are placed a) all with one of the opponents, b) three with one opponent and one with the other, c) two with each of the opponents.

14. Find the probability that, at a deal for bridge, at least one of the players will obtain exclusively cards of the values 2, 3, . . ., 9.

15. Find the probability for a poker player to have "full house" in the first round. — It is assumed that the game is played with an ordinary set of cards from which all cards of the values 2, 3, . . ., 6 have been discarded, while a joker has been added. In the first round five cards are dealt to each player, and "full house" consists of three cards of one value and two of another value. Such a combination as, e.g., three aces, one king, and the joker is, however, not counted as "full house" but as "four of a kind", which is a higher combination.

16. A sequence of independent repetitions of a random experiment is performed. In each repetition the probability of a "success" is equal to p. Find the probability that exactly n repetitions will be required in order to reach a given number ν of successes.

17. n throws are made with an ordinary die. Find the probability that a) the greatest, b) the smallest, of the numbers obtained will have a given value k.

18. An urn contains balls of four different colours, each colour being represented by the same number of balls. Four balls are drawn with replacement. Find the probability that at least three different colours will appear in the drawings.

19. Three players A, B, and C of equal skill are engaged in a sequence of games, where the winner of each game scores one point. The one that first scores three points will be the final winner. A wins the first and the third game, while B wins the second. Find the probability that C will be the final winner.

20. A table of random sampling numbers is a table designed to give a sequence of digits that might have been obtained by repeated drawings with replacement from an urn containing ten similar balls marked $0, 1, \ldots, 9$. Find the probability that in a sequence of ten digits from such a table all ten digits will actually appear.

21. Each package of cigarrettes of a certain brand contains a card bearing one of the numbers $1, 2, \ldots, k$. Among the very large number of packages which are in the market all k numbers occur equally often. A person buys n packages. Find the probability that he will have at least one complete set of all k numbers.

22. The urn U_1 contains N balls, one of which is black, while the others are white. U_2 contains N balls, all of which are white. One ball is drawn from each urn and placed in the other urn, and this procedure is repeated a number of times. Find the probability that after n repetitions the black ball will be in U_1. Show that this probability tends to a definite limit as n tends to infinity.

23. Each of the two urns U_1 and U_2 contains one white and one black ball. One ball is drawn from each urn and placed in the other urn, and this procedure is repeated a number of times. Let p_n, q_n, and r_n denote the probabilities that, after n repetitions, the urn U_1 will contain two white balls, one white and one black, or two black balls respectively. Deduce formulas expressing p_{n+1}, q_{n+1}, and r_{n+1} in terms of p_n, q_n, and r_n. Use these to find the probabilities p_n, q_n, and r_n, and show that these quantities tend to definite limiting values as n tends to infinity.

24. Suppose that for each day the weather can be uniquely classified as "fine" or "bad". Suppose further that the probability of having fine weather on the last day of a certain year is P_0, and that we have the probability p that the weather on an arbitrary day will be of the same kind as on the preceding day. Find the probability P_n of having fine weather on the n:th day of the following year. Find the limiting value of P_n as n tends to infinity.

25. Each of the urns $U_0, U_1, \ldots, U_N$ contains N balls. U_i contains i white and $N - i$ black balls. We select one urn at random, and from this we draw n balls with replacement. Among these we find ν white and $n - \nu$ black balls. a) Find the probability *a posteriori* that we have selected the urn U_i. b) One more ball is drawn from the same urn. Find the probability that this ball will be white, and study the behaviour of this probability as N tends to infinity.

PART II

RANDOM VARIABLES AND
PROBABILITY DISTRIBUTIONS

CHAPTER 5

VARIABLES AND DISTRIBUTIONS IN ONE DIMENSION

5.1. Introduction. — The concept of a random experiment has been introduced in 2.2. In most cases the result of an individual performance of a random experiment will be expressed in numerical form, i.e., by stating the values assumed in the experiment by a certain number of variable quantities. In the first place, this is obviously true whenever the experiment is directly concerned with the counting or measurement of some quantitative variables. By means of appropriate conventions we may, however, use the numerical mode of expression also in many other cases. If, e.g., the experiment consists in observing the sex of a newborn child, we may agree to denote "boy" by 1, and "girl" by 0. Similar conventions may be used also in more complicated cases of observations on non-quantitative variables.

Until further notice we shall deal exclusively with cases where the result of a random experiment is expressed by means of *one single* numerical quantity.

A variable quantity which expresses the result of a given random experiment will be called a *random* or *stochastic variable*. If X denotes such a variable, X will in general assume different values in different performances of the random experiment to which X is attached. An *observed value* of X is a value that has been observed at an actual performance of the experiment.

By $P(X=a)$ we denote the probability of the event that the variable X assumes the given value a. Similarly $P(a<X\leq b)$ denotes the probability of the event that X assumes a value belonging to the interval $a<X\leq b$, etc. If we knew the probability $P(a<X\leq b)$ for all values of a and b, we should have a complete survey of the probabilities with which the variable tends to assume values in different parts of the scale of numbers. In a case when this is true, we shall accordingly say that we know the *probability distribution*, or briefly the *distribution*, of the variable X.

Let now x denote a given number, and consider the probability $P(X\leq x)$ that the variable X will assume a value such that $X\leq x$. Obviously this probability will be a function of x. Let us write

(5.1.1) $$F(x)=P(X\leq x),$$

57

and call $F(x)$ the *distribution function* (abbreviated *d.f.*) of the random variable X. Let now a and b be any real numbers such that $a < b$. The events $X \leq a$ and $a < X \leq b$ will then be mutually exclusive, and their sum as defined in 3.2 will be the event $X \leq b$. By means of the addition rule 3.2.1 a it then follows that we have

$$P(X \leq b) = P(X \leq a) + P(a < X \leq b),$$

and hence

(5.1.2) $$P(a < X \leq b) = F(b) - F(a).$$

If we know the distribution function $F(x)$ for all values of x, the probability distribution of the variable will thus be completely known.

Since every probability is a non-negative number, it follows from 5.1.2 that $F(b) \geq F(a)$. Any distribution function $F(x)$ will thus be a *never decreasing* function of the real variable x.

Confining ourselves to variables assuming only finite values. we find

(5.1.3) $$F(+\infty) = \lim_{x \to \infty} F(x) = 1,$$

since this limiting value coincides with the probability that the variable will assume *any* finite value. In a similar way we obtain

(5.1.4) $$F(-\infty) = \lim_{x \to -\infty} F(x) = 0.$$

If in 5.1.2 we first keep b fixed and allow a to tend to b, and then perform the opposite limiting process, we further find

$$F(a) - F(a-0) = P(X = a),$$

(5.1.5)

$$F(a+0) - F(a) = 0.$$

The second relation in 5.1.5 shows that any distribution function $F(x)$ is *continuous to the right* for every value of x. On the other hand, the first relation in 5.1.5 shows that $F(x)$ is *discontinuous to the left* in each point $x = a$ such that $P(X = a) > 0$, and that in such a point $F(x)$ has a step of the height $P(X = a)$. Conversely, it follows from the first relation in 5.1.5 that if $F(x)$ is continuous in a certain point $x = a$, then $P(X = a) = 0$.

We now introduce a convenient illustration which we shall very often have the occasion to use. Let us conceive the axis of X as an infinitely thin bar over which a *mass* of the total quantity 1 is distributed. The density of the mass may vary between different parts of the bar. Some parts of the bar may contain no mass at all, while on the other hand certain isolated points of the bar may be bearers of positive mass quantities concentrated in these points. If we denote by $F(x)$ the amount of mass which is situated to the left of the point x, or in the point x itself, we shall find that the function $F(x)$ will have exactly the same properties as those above deduced for a distribution function The quantity of mass belonging to an arbitrary interval $a < X \leq b$ will be equal to $F(b) - F(a)$, and will thus correspond to the probability that the variable assumes a value belonging to this interval. By way of illustration, we may thus say that any distribution function defines a *distribution of mass* over the axis of X, and talk of a *probability mass* which is distributed over the axis according to the d.f. $F(x)$. This will enable us to use certain well-known concepts and properties belonging to the field of elementary mechanics in order to illustrate the definitions and propositions of probability theory.

As we have seen, any probability distribution is completely characterized by means of the corresponding distribution function $F(x)$. For a general mathematical theory of probability distributions, $F(x)$ will thus be the natural instrument. However, the great majority of probability distributions occurring in practical applications belong to the one or the other of two simple particular types known as the *discrete* and the *continuous* type. For these types, there are simpler and more intuitive means of describing the distribution, as we shall see in the two following sections.

5.2. Discrete distributions. — A random variable and its probability distribution will be said to belong to the *discrete type* if, in terms of the mechanical illustration discussed in the preceding section, the total mass of the distribution is concentrated in certain isolated points. Moreover, we shall always impose the condition that in an arbitrary finite interval there shall be at most a finite number of these mass points.

Let $x_1, x_2, \ldots$ be the finite or infinite sequence of mass points, while $p_1, p_2, \ldots$ are the corresponding quantities of mass. Then $x_1, x_2, \ldots$ are the only possible values of the random variable X, and the probability that X will assume a given value x_ν is equal to p_ν:

$$P(X = x_\nu) = p_\nu \quad (\nu = 1, 2, \ldots).$$

Since the total mass in the distribution must be unity, we have

(5.2.1) $$\sum_\nu p_\nu = 1,$$

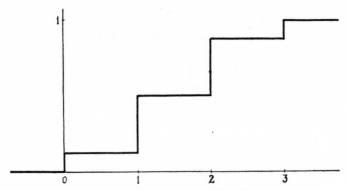

Fig. 2. Distribution function of the discrete type.

Fig. 3. Probability diagram for the distribution shown in fig. 2.

where the sum should be extended over the whole finite or infinite sequence of mass points.

The distribution function $F(x)$ for a distribution of this type will be

$$F(x) = \sum_{x_\nu \leq x} p_\nu,$$

where the sum should be extended over all mass points x_ν which are $\leq x$. Thus a distribution function of this type is a "step function" which is constant over every interval not containing any of the mass points x_ν. In each x_ν the distribution function has a discontinuity with a "step" of the height p_ν.

An important particular case when distribution functions of the discrete type are encountered in practice arises whenever we are concerned with a random variable which denotes the *number* of certain objects or individuals. If, e.g., the random variable is the absolute frequency f of a certain event A in a sequence of n independent repetitions of a given random experiment, the possible values of the variable f will be $0, 1, \ldots, n$. If the probability of the occurrence of A in any repetition is denoted by p, we have according to 4.1.1

$$p_\nu = P(f = \nu) = \binom{n}{\nu} p^\nu q^{n-\nu}.$$

In each point $x = \nu$, where $\nu = 0, 1, \ldots, n$, the corresponding distribution function has a discontinuity with a step of the height p_ν. In every interval not containing any of the points ν, the function is constant. In fig. 2 this function is graphically represented for the case $n = 3$, $p = \frac{1}{2}$, where we have $p_0 = p_3 = \frac{1}{8}$ and $p_1 = p_2 = \frac{3}{8}$, while the corresponding variable may be, e.g., the number of heads in three throws with a coin.

A more intuitive way of illustrating a distribution of this type is simply to erect over every point x_ν an ordinate equal to the corresponding probability p_ν. By 5.2.1, the sum of all these ordinates will then always be unity. In fig. 3, this *probability diagram* is drawn for the case of the special distribution just mentioned.

5.3. Continuous distributions. — A random variable and its probability distribution will be said to belong to the *continuous type* if the total mass in the corresponding mass distribution is continuously distributed, with a density $f(x)$ which is everywhere continuous, except possibly in certain isolated discontinuity points, of which there are at most a finite number in an arbitrary finite interval. Obviously we shall always have $f(x) \geqq 0$ for every value of x.

The amount of mass in an infinitesimal interval $(x, x + dx)$ will then be $f(x)\,dx$, which corresponds to the infinitesimal probability that the variable will assume a value belonging to this interval. Thus the density $f(x)$ will be a measure of the frequency with which, in a sequence of repeated experiments, the variable will tend to assume a value belonging to this interval, and accordingly the function $f(x)$ will be called the *probability density* (abbreviated *pr. d.*) or the *frequency function* (abbreviated *fr. f.*) of the distribution. The differential $f(x)\,dx$ is known as the *probability element* of the distribution.

The amount of mass in an arbitrary interval $a < X \leq b$, which corresponds to the probability that the variable X will assume a value belonging to this interval, will be

$$(5.3.1) \qquad P(a < X \leq b) = F(b) - F(a) = \int_a^b f(x)\,dx.$$

If, in particular, we take here $a = -\infty$, we obtain

$$(5.3.2) \qquad F(b) = \int_{-\infty}^b f(x)\,dx,$$

and for $b = +\infty$

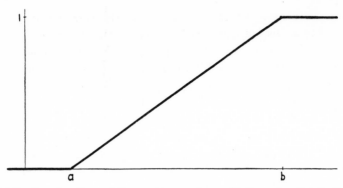

Fig. 4. Distribution function of the continuous type. Rectangular distribution over the interval (a, b).

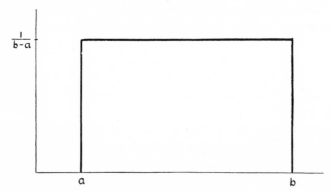

Fig. 5. Frequency function for the distribution shown in fig. 4.

$$(5.3.3) \qquad \int\limits_{-\infty}^{\infty} f(x)\,dx = 1,$$

which means that the total mass of the distribution is unity. On the other hand, we obtain by differentiation of 5.3.2

$$F'(x) = f(x).$$

Thus the frequency function is the derivative of the distribution function. It will be seen that any function $f(x)$, which has the continuity properties stated above, is non-negative for all values of x, and satisfies 5.3.3, defines a distribution of the continuous type, having $f(x)$ for its frequency function. Since a distribution function of this type is everywhere continuous, we have according to 5.1 for every value of a

$$P(X = a) = 0.$$

62

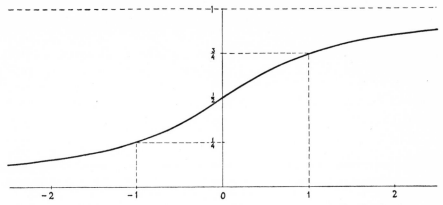

Fig. 6. Distribution function of the continuous type. CAUCHY's distribution.

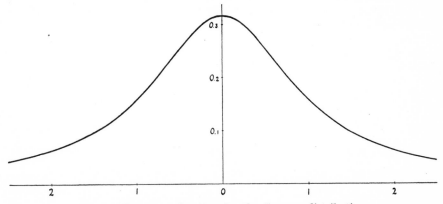

Fig. 7. Frequency function for the CAUCHY distribution.

Variables of the continuous type are encountered whenever we are concerned with a random experiment which consists in the measurement of quantities capable of assuming any value within certain limits. Examples are the measurement of the stature of a man, the sugar level in human blood, the duration of a telephone conversation, the atomic weight of a chemical element, etc.

A distribution of the continuous type can be graphically illustrated by means of a diagram showing the *distribution curve* $y = F(x)$ or, more intuitively, the *frequency curve* $y = f(x)$. For the particular distributions considered in the two examples given below, such diagrams are shown in fig. 4—7.

The total area situated between a frequency curve $y = f(x)$ and the axis of x is by 5.3.3 always equal to unity. The section of this area included between the ordinates through the points $x=a$ and $x=b$ is given by the integral $\int_a^b f(x)\,dx$, and by 5.3.1 this area is thus equal to the probability that the corresponding random variable X will assume a value belonging to the interval (a, b). This probability will have the same value whether the extreme points a and b are included in the interval or not, since for a continuous distribution the probabilities $P(X=a)$ and $P(X=b)$ are both zero.

If $y = F(x)$ is the distribution curve of a continuous distribution, and if ε is a given arbitrarily small positive quantity, it is readily seen that we can always find a distribution function $G(x)$ belonging to a discrete distribution such that $\left| F(x) - G(x) \right| < \varepsilon$ for all x. In fact, we have only to draw a "step curve" $y = G(x)$ such that the two curves come sufficiently close together. In this way, any continuous distribution can be arbitrarily closely approximated by a discrete distribution.

We shall now give two simple examples of continuous distributions. Further examples will be studied in chapters 7–8.

Ex. 1. — A random variable X with the frequency function $f(x)$ defined by the relations

$$f(x) = \begin{cases} \dfrac{1}{b-a} & \text{for } a < x < b, \\ 0 & \text{otherwise,} \end{cases}$$

will be said to have a *rectangular distribution* in the interval (a, b), or to be *uniformly distributed* over (a, b). The corresponding distribution function is equal to zero for $x < a$, equal to unity for $x \geq b$, and

$$F(x) = \frac{x-a}{b-a}$$

for $a < x < b$. Diagrams of $F(x)$ and $f(x)$ are shown in fig, 4–5.

Ex. 2. The distribution defined by the frequency function

$$f(x) = \frac{1}{\pi(1+x^2)}$$

for all real values of x, is known as CAUCHY's distribution. The corresponding distribution function $F(x)$ is defined according to 5.3.2. Diagrams of $F(x)$ and $f(x)$ are shown in fig. 6–7.

Suppose now that X is a random variable with the d.f. $F(x)$, and consider a function $Y = \varphi(X)$. Let us modify the definition of the random experiment to which the variable X is attached by prescribing that we have 1) to observe the value assumed by X in the original experiment, 2) to compute the corresponding value $Y = \varphi(X)$, and 3) to present the value Y as the result of the modified experiment. If $\varphi(X)$ is a simple function, the probability distribution of the new random variable $Y = \varphi(X)$ attached to the modified experiment can then often be simply expressed in terms of the given $F(x)$.

We shall here only consider the simple case when X is a variable belonging to the continuous type, while Y is a linear function $Y = \varphi(X) = a X + b$. The relation $Y \leq y$ will then be equivalent to $X \leq \dfrac{y - b}{a}$, or to $X \geq \dfrac{y - b}{a}$, according as $a > 0$ or $a < 0$. Thus if we denote the d.f. of Y by $G(y)$, and the corresponding fr. f. by $g(y) = G'(y)$, we shall have

$$(5.3.4) \qquad G(y) = P(Y \leq y) = \begin{cases} F\left(\dfrac{y-b}{a}\right) & \text{for } a > 0, \\[2ex] 1 - F\left(\dfrac{y-b}{a}\right) & \text{for } a < 0, \end{cases}$$

and in both cases

$$(5.3.5) \qquad g(y) = \frac{1}{|a|} f\left(\frac{y-b}{a}\right).$$

5.4. Joint distributions of several variables. — In many cases the result of a random experiment will not be expressed by one single observed quantity, but by a certain number of simultaneously observed quantities. Thus, e.g., we may measure the stature and the weight of a man, the hardness and the percentage of carbon of a sample of steel, etc. A detailed discussion of questions occurring in this connection must be postponed until chapters 9—10; already at this stage it will, however, be necessary to introduce certain fundamental notions that will be required in the following sections.

If the result of a certain random experiment is expressed by means of two simultaneously observed, variable quantities X and Y, and if we represent X and Y as rectangular coordinates in a certain plane, every pair of simultaneously observed values of the two random variables X and Y will correspond to a certain point in the plane. By

$$P(a_1 < X \leq b_1, \ a_2 < Y \leq b_2)$$

we denote the probability that this point will belong to the rectangle defined by the inequalities within the brackets. If we know this probability for all values of a_1, b_1, a_2, and b_2, we shall say that we know the *joint* or *simultaneous distribution* of the random variables X and Y.

We now define the *joint distribution function* of X and Y by writing

$$(5.4.1) \qquad F(x, y) = P(X \leq x, Y \leq y).$$

By similar arguments as in the case of 5.1.2, we then obtain

$$P(a_1 < X \leq b_1, a_2 < Y \leq b_2) = F(b_1, b_2) - F(a_1, b_2) - F(b_1, a_1) + F(a_1, a_2),$$

which shows that, as in the case of one single variable, the distribution function gives a complete characterization of the probability distribution. In the same way as in 5.1, we further see that the distribution function is a never decreasing function of each of its two arguments, and that $F(x, y)$ tends to 1 as both arguments tend simultaneously to $+\infty$, while $F(x, y)$ tends to 0 as one of the two arguments tends to $-\infty$, while the other argument is kept constant.

In a similar way as in the case of one single variable, the *two-dimensional distribution* considered here can be illustrated by means of a mechanical analogy. In the present case we have to imagine a unit of mass distributed over the (X, Y)-plane, and the mass within an arbitrary rectangle $a_1 < X \leq b_1$, $a_2 < Y \leq b_2$ will then correspond to the probability that the point (X, Y) belongs to this rectangle. The distribution function $F(x, y)$ is equal to the amount of mass in the infinitely extended "rectangle" $X \leq x$, $Y \leq y$.

As before it will be useful to consider in particular two simple types of distributions: the discrete and the continuous type.

In a distribution of the *discrete type*, each variable has only a finite or infinite sequence of isolated possible values. If $x_1, x_2, \ldots$ and $y_1, y_2, \ldots$ are the possible values of X and Y respectively, we denote by

$$p_{\mu\nu} = P(X = x_\mu, Y = y_\nu), \qquad (\mu, \nu = 1, 2, \ldots)$$

the probability that simultaneously X will assume the value x_μ and Y the value y_ν. The total mass in the distribution will then be

$$\sum_{\mu, \nu} p_{\mu\nu} = 1.$$

We further denote by p_μ. the probability that X assumes the value x_μ, without paying regard to the value assumed by Y. For two different pairs of subscripts μ, ν, the event defined by the simultaneous relations

$X = x_\mu$, $Y = y_\nu$, will always be mutually exclusive, and thus we find by the aid of the addition rule 3.2.3 a[1]

$$(5.4.2) \qquad\qquad p_\mu. = \sum_\nu p_{\mu\nu},$$

where the sum should be extended over all values of ν. Thus if we completely disregard the values assumed by Y, and consider X by itself as a random variable connected with our given experiment, X will have a distribution of the discrete type, with the possible values x_μ and the corresponding probabilities p_μ. This one-dimensional distribution will be called the *marginal distribution of* X attached to the given joint distribution of X and Y. In a similar way we denote by $p_{.\nu}$ the probability that Y assumes the value y_ν, without regard to the value assumed by X, and we then find

$$(5.4.3) \qquad\qquad p_{.\nu} = \sum_\mu p_{\mu\nu}.$$

Thus also the *marginal distribution of* Y belongs to the discrete type, having the possible values y_ν, and the corresponding probabilities $p_{.\nu}$.

For a distribution of the *continuous type*, there exists a *probability density* or *frequency function* $f(x, y)$ which, in the mechanical illustration, represents the density of the mass in the point (x, y), and which is everywhere finite and continuous, except possibly in certain lines or curves in the plane, where discontinuities may occur. The probability that the variables X and Y assume values within the infinitesimal rectangle $x < X < x + dx$, $y < Y < y + dy$, will then be the *probability element* of the distribution:

$$f(x, y)\, dx\, dy.$$

The amount of probability mass belonging to an arbitrary rectangle $a_1 < X \leqq b_1$, $a_2 < Y \leqq b_2$, is

$$P(a_1 < X \leqq b_1, \ a_2 < Y \leqq b_2) = \int\limits_{a_1}^{b_1} \int\limits_{a_2}^{b_2} f(x, y)\, dx\, dy,$$

and for the total mass in the whole plane we find

$$\int\limits_{-\infty}^{\infty} \int\limits_{-\infty}^{\infty} f(x, y)\, dx\, dy = 1.$$

---

[1] If there are only a finite number of values of ν, this follows directly from 3.2.3 a. For an infinite sequence, the same formula is obtained by a simple passage to the limit.

The probability that $X \leq x$, without regard to the value assumed by Y, will be denoted by $F_1(x)$. We then have

$$F_1(x) = P(X \leq x) = \int_{-\infty}^{x} du \int_{-\infty}^{\infty} f(u, y)\, dy.$$

Obviously this will be the distribution function associated with the *marginal distribution of* X. For all distributions occurring in the applications, this will be a distribution function of the continuous type, and the corresponding frequency function is found by differentiation:

$$(5.4.4) \qquad\qquad f_1(x) = F_1'(x) = \int_{-\infty}^{\infty} f(x, y)\, dy.$$

For the distribution function and the frequency function associated with the *marginal distribution of* Y, we find in the same way the expressions

$$F_2(y) = \int_{-\infty}^{y} dv \int_{-\infty}^{\infty} f(x, v)\, dx,$$

$$(5.4.5)$$

$$f_2(y) = F_2'(y) = \int_{-\infty}^{\infty} f(x, y)\, dx.$$

The definition of the marginal distributions of X and Y can be simply illustrated by means of the mechanical analogy. Starting from the two-dimensional mass distribution corresponding to the joint distribution of X and Y, let us imagine that every mass particle moves orthogonally towards the axis of X, until it arrives in a point of this axis. By this flow of mass in the plane, the whole two-dimensional distribution is projected on the axis of X, thus generating a one-dimensional mass distribution on this axis which will obviously correspond to the marginal distribution of X. In the same way the marginal distribution of Y is obtained by projecting the two-dimensional distribution on the axis of Y.

If new variables X' and Y' are introduced by a linear substitution

$$X = \alpha X' + \beta Y',$$

$$Y = \gamma X' + \delta Y'.$$

it follows from well-known propositions in the integral calculus that the probability element $f(x, y)\, dx\, dy$ is transformed according to the formula

$$f(x, y)\, dx\, dy = |\alpha \delta - \beta \gamma|\, f(\alpha x' + \beta y', \gamma x' + \delta y')\, dx'\, dy'.$$

The joint frequency function of the new variables X' and Y' will thus be

(5.4.6) $$|\alpha\delta - \beta\gamma|\, f(\alpha x' + \beta y',\, \gamma x' + \delta y').$$

Let us now consider two variables X and Y, having a joint distribution of arbitrary type. Consider the two events defined by the relations $a_1 < X \leq b_1$ and $a_2 < Y \leq b_2$. If, for all values of a_1, b_1, a_2 and b_2, these are independent events (3.5), we shall say that X and Y are *independent random variables*. This concept will find many important applications in the sequel. For two independent variables, we have according to 3.5.1

$$P(a_1 < X \leq b_1,\ a_2 < Y \leq b_2) = P(a_1 < X \leq b_1)\, P(a_2 < Y \leq b_2),$$

and hence in particular, as a_1 and a_2 tend to $-\infty$,

(5.4.7) $$F(x,\,y) = F_1(x)\, F_2(y),$$

where $F_1(x)$ and $F_2(y)$ are the distribution functions of the two marginal distributions. If the distribution is of the discrete type we further have, using the notations introduced above,

(5.4.8) $$p_{\mu\nu} = p_{\mu\cdot}\, p_{\cdot\nu}.$$

On the other hand, for a distribution of the continuous type we find, taking $a_1 = x$ and $b_1 = x + dx$, and similarly for y,

(5.4.9) $$f(x,\,y) = f_1(x)\, f_2(y).$$

Conversely it is easily seen that, as soon as one of the three last relations holds for all values of x and y (or μ and ν), it follows that the two variables are independent. Any of the three relations 5.4.7–5.4.9 thus constitutes a necessary and sufficient condition for the independence of the random variables X and Y.

Ex. 1. — Consider a distribution of two variables X and Y, where each variable only has the possible values 0 and 1. This is a distribution of the discrete type, where the total mass is concentrated in the four points (0,0), (0,1) (1,0) and (1,1). Let the masses in these points be $p_{00} = p_{11} = \frac{1}{2} p$, $p_{01} = p_{10} = \frac{1}{2} q$, where p and q are non-negative numbers such that $p + q = 1$. The marginal distribution of X will then have the possible values 0 and 1, with the probabilities $p_{0\cdot} = p_{00} + p_{01} = \frac{1}{2}$, and $p_{1\cdot} = p_{10} + p_{11} = \frac{1}{2}$. Similarly, the marginal distribution of Y will have the possible values 0 and 1, with the probabilities $p_{\cdot 0} = p_{\cdot 1} = \frac{1}{2}$. If, in particular, $p = q = \frac{1}{2}$, it follows from 5.4.8 that the variables in this distribution are independent. As soon as $p \neq \frac{1}{2}$, however, we shall e.g. have $p_{00} \neq p_{0\cdot}\, p_{\cdot 0}$, so that the condition of independence will not be satisfied.

Ex. 2. — We define a distribution of the continuous type by taking

$$f(x, y) = \frac{1}{(b_1 - a_1)(b_2 - a_2)}$$

for $a_1 < x < b_1$, $a_2 < y < b_2$, and $f(x, y) = 0$ otherwise. The mass in this distribution is distributed with a constant density over the area of the rectangle $a_1 < x < b_1$, $a_2 < y < b_2$. Show by the aid of 5.4.4 that the marginal distribution of X is a rectangular distribution (5.3, ex. 1) over the interval $a_1 < x < b_1$, and similarly for Y! Further, use 5.4.9 to show that the variables in this distribution are independent!

Finally we shall very briefly consider the generalization of the preceding discussion to cases involving joint distributions of more than two variables.

If $X_1, X_2, \ldots, X_n$ are random variables, all of which are simultaneously observed in the same random experiment, we define

$$(5.4.10) \qquad F(x_1, \ldots, x_n) = P(X_1 \leqq x_1, \ldots, X_n \leqq x_n)$$

as the joint distribution function of the variables. The joint probability distribution of the variables will, in the usual way, have its mechanical analogy in a distribution of a unit of mass over the n-dimensional space with the coordinates $X_1, \ldots, X_n$. In a distribution of the *discrete type*, the total mass is concentrated in certain isolated points, and the distribution will be completely characterized by the finite or infinite sequence of these points, together with the corresponding probabilities. On the other hand, a distribution of the *continuous type* is characterized by the *probability density* or the *frequency function* $f(x_1, \ldots, x_n)$, where the *probability element*

$$f(x_1 \ldots, x_n)\, dx_1 \ldots dx_n$$

expresses the probability that we shall have $x_\nu < X_\nu \leqq x_\nu + dx_\nu$ for all $\nu = 1, 2, \ldots, n$. The variables in a distribution of arbitrary type are *independent* if

$$F(x_1, \ldots, x_n) = F_1(x_1) \ldots F_n(x_n),$$

where $F_\nu(x_\nu)$ denotes the marginal distribution function of X_ν, i.e., the distribution function of X_ν, without regard to the values assumed by the other variables. If the distribution belongs to the continuous type, the independence condition can also be written

$$(5.4.11) \qquad f(x_1, \ldots, x_n) = f_1(x_1) \ldots f_n(x_n),$$

where $f_\nu(x_\nu)$ is the marginal frequency function of X_ν, i.e., the frequency function corresponding to the distribution function $F_\nu(x_\nu)$.

5.5. Mean values. — Let X be a random variable with a distribution of arbitrary type, and let us as before consider the corresponding distribution of mass. The abscissa of the *centre of gravity* of the mass in this distribution constitutes a weighted mean value of the various values capable of being assumed by the variable, with the corresponding mass particles or probabilities as weights. This value will be called the *mean value*, or briefly the *mean*, of the variable X, or of the corresponding distribution, and will be denoted by $E(X)$ or, when there is no risk of misunderstanding, briefly by $E\,X$.

From well-known propositions in elementary mechanics, it then follows for a distribution of the discrete type[1]

$$(5.5.1) \qquad E(X) = \sum_{\nu} p_{\nu}\, x_{\nu},$$

and for a distribution of the continuous type

$$(5.5.2) \qquad E(X) = \int_{-\infty}^{\infty} x\, f(x)\, dx.$$

We shall say that the mean value *exists*, only when the corresponding sum or integral is *absolutely convergent*. This will, e.g., always be the case when the number of terms in 5.5.1 is finite, or when the frequency function $f(x)$ in 5.5.2 vanishes everywhere outside a certain finite interval. As soon as these conditions are not satisfied, it will be necessary to ascertain the convergence of the series or the integral.

According to 5.3, a function $Y = \varphi(X)$ is a random variable which has a certain distribution, and a certain mean value determined by this distribution. If the distribution of X belongs to one of the two simple types, this mean value can be expressed by means of formulae, which are a direct generalization of 5.5.1 and 5.5.2 (cf. *Random Variables* p. 19):

---

[1] The reader not sufficiently acquainted with these mechanical concepts may simply take the formulae 5.5.1 and 5.5.2 as definitions of the mean value $E(X)$. With the aid of the generalized integral concept known as the STIELTJES integral, these formulae can be considered as particular cases of a general formula which expresses the mean value for an arbitrary distribution. However, a discussion of the important questions belong to this order of ideas falls outside the scope of this book, and we must content ourselves with referring to more extensive treatises. — We finally remark that the choice of the letter E to denote a mean value is due to the circumstance that the concept of mean value was first introduced in the theory of games, where the mean value of the gain of a player can be interpreted as his mathematical *expectation*. See further 6.3.

$$E\left(\varphi\left(X\right)\right)=\sum_{\nu}p_{\nu}\,\varphi\left(x_{\nu}\right),$$

(5.5.3)

$$E\left(\varphi\left(X\right)\right)=\int_{-\infty}^{\infty}\varphi\left(x\right)f\left(x\right)dx.$$

Similar formulae hold for functions of more than one random variable. Thus for a function $Z=\varphi\left(X,Y\right)$ of two variables we have, according as the joint distribution of X and Y is of the discrete or the continuous type (5.4)

$$E\left(\varphi\left(X,\,Y\right)\right)=\sum_{\mu,\nu}p_{\mu\nu}\,\varphi\left(x_{\mu},\,y_{\nu}\right),$$

(5.5.4)

$$E\left(\varphi\left(X,\,Y\right)\right)=\int_{-\infty}^{\infty}\int_{-\infty}^{\infty}\varphi\left(x,y\right)f\left(x,y\right)dx\,dy.$$

In all these cases we shall say as before that a mean value *exists*, only if the corresponding sum or integral is *absolutely convergent*.

Ex. 1. — Let X denote the number of points obtained in a throw with an ordinary die. By 5.5.1 we then have $E\left(X\right)=\dfrac{1}{6}(1+2+\cdots+6)=\dfrac{7}{2}$. Thus it is not necessary that the mean value coincides with one of the possible values of the variable.

Ex. 2. — If X has a rectangular distribution (5.3, ex. 1) in (a,b), we obtain from 5.5.2 the mean value $E\left(X\right)=\dfrac{1}{b-a}\int_{b}^{a}x\,dx=\dfrac{a+b}{2}.$

Ex. 3. — For the CAUCHY distribution (5.3, ex. 2) we obtain from 5.5.2

$$E\left(X\right)=\frac{1}{\pi}\int_{-\infty}^{\infty}\frac{x}{1+x^{2}}\,dx.$$

However, it is easily seen that this integral is not absolutely convergent, so that we must say that, for this distribution, no mean value exists. If, instead of the mean value of the variable X, we consider the mean value of the function $Y=\varphi\left(X\right)=\left|X\right|^{\alpha}$, where $0<\alpha<1$, then 5.5.3 will give an absolutely convergent integral, so that this mean value exists.

We now proceed to deduce some general rules for the calculation with mean values. The relation 5.5.3 gives, in the particular case when $\varphi\left(X\right)$ is linear,

(5.5.5)
$$E\left(a\,X+b\right)=a\,E\left(X\right)+b.$$

If, in 5.5.4, we take $\varphi(X, Y) = X + Y$ and use 5.4.2 – 5.4.5 we obtain, according as the joint distribution is discrete or continuous,

$$E(X + Y) = \sum_{\mu,\nu} p_{\mu\nu} x_\mu y_\nu = \sum_\mu p_\mu . x_\mu + \sum_\nu p_{.\nu} y_\nu,$$

$$E(X + Y) = \int_{-\infty}^{\infty} \int_{-\infty}^{\infty} (x + y) f(x, y) \, dx \, dy = \int_{-\infty}^{\infty} x f_1(x) \, dx + \int_{-\infty}^{\infty} y f_2(y) \, dy.$$

In both cases we have here expressed $E(X + Y)$ as a sum of two terms, which are identical with the mean values of the variables X and Y in their respective marginal distributions. In both cases we shall thus have

$$(5.5.6) \qquad E(X + Y) = E(X) + E(Y).$$

Replacing here Y be $Y + Z$, we obtain a corresponding addition formula for three variables, and by a repeated application of the same procedure we obtain the important general formula

$$(5.5.7) \qquad E(X_1 + X_2 + \cdots + X_n) = E(X_1) + E(X_2) + \cdots + E(X_n).$$

This formula constitutes the general *addition rule for mean values*. We have here proved this rule only for distributions belonging to one of the two simple types, but in reality the formula holds for all kinds of distributions. It is particularly important to note that, as our deduction of the addition formulae 5.5.6 and 5.5.7 shows, these formulae hold good *whether the variables are independent or not*.

We now proceed to deduce a similar rule for the mean value of a *product*, and it will then be necessary to introduce considerably more restrictive assumptions. Let us assume that X and Y are two *independent* random variables (as shown by 9.5, ex. 1, this assumption could, in fact, be somewhat generalized). If the joint distribution is of the continuous type, we have by 5.4.9 $f(x, y) = f_1(x) f_2(y)$, and it thus follows

$$(5.5.8) \qquad \begin{aligned} E(XY) &= \int_{-\infty}^{\infty} \int_{-\infty}^{\infty} x y f(xy) \, dx \, dy = \int_{-\infty}^{\infty} x f_1(x) \, dx \int_{-\infty}^{\infty} y f_2(y) \, dy \\ &= E(X) E(Y). \end{aligned}$$

When the joint distribution is of the discrete type, the same formula can be proved by means of 5.4.8, and it will be left to the reader to work out this proof in detail. In the same way as the addition rule, this *multiplication rule* can be generalized to the case of more than two variables. If $X_1, \ldots, X_n$ are *independent* variables, we thus have

$$(5.5.9) \qquad E(X_1 \ldots X_n) = E(X_1) \ldots E(X_n).$$

Like the addition rules 5.5.6 and 5.5.7, the multiplication rules 5.5.8 and 5.5.9 hold, in fact, not only for distributions belonging to the two simple types, but for completely general distributions. The only condition for the validity of all these formulae is that all the occurring mean values should exist, and further, in respect of the multiplication rules 5.5.8 and 5.5.9, that the variables should be independent.

When, in the sequel, we are concerned with mean values, we shall always tacitly assume that all mean values involved in our calculations exist. Any result obtained in this way will then hold subject to the obvious condition that this assumption is satisfied.

5.6. Moments. — The mean value of the power X^ν, where ν is a positive integer, will be called the ν:th *moment*, or the moment of order ν, of the variable X, or of the corresponding distribution. If we write

$$(5.6.1) \qquad \alpha_\nu = E(X^\nu),$$

we shall obviously always have $\alpha_0 = 1$, while α_1 will be identical with the mean value of the variable, which we shall often denote by the letter m:

$$\alpha_1 = E(X) = m.$$

More generally, the mean value of the power $(X-c)^\nu$, where c is a constant, will be called the moment of order ν *about the point c*. The moments α_ν are thus the moments about the origin.

The moments about the mean value m are particularly important. We shall write

$$(5.6.2) \qquad \mu_\nu = E((X-m)^\nu),$$

and call μ_ν the *central moment* of order ν. By 5.5.5 we shall then always have

$$(5.6.3) \qquad \begin{aligned} \mu_1 &= E(X) - m = 0, \\ \mu_2 &= E(X^2) - 2m\,E(X) + m^2 = \alpha_2 - m^2. \end{aligned}$$

In the same way we obtain expressions for the higher central moments. We shall here only give the formulae for μ_3 and μ_4:

$$(5.6.4) \qquad \begin{aligned} \mu_3 &= \alpha_3 - 3\,m\,\alpha_2 + 2\,m^3, \\ \mu_4 &= \alpha_4 - 4\,m\,\alpha_3 + 6\,m^2\,\alpha_2 - 3\,m^4. \end{aligned}$$

Thus the central moments can easily be computed with the aid of the moments about the origin. Moreover, it is easily seen that, by means of similar formulae, the moments about an arbitrary point c_1 can always be expressed in terms of the moments about any other given point c_2.

The central moment of the second order, $\mu_2 = E((X-m)^2)$, will play an important part in the sequel. Since μ_2 is the mean value of a variable $(X-m)^2$ which can never assume negative values, we always have $\mu_2 \geq 0$. Further, for an arbitrary point $c \neq m$ we have

$$E((X-c)^2) = E((X-m+m-c)^2)$$
$$= E((X-m)^2) - 2(c-m)E(X-m) + (c-m)^2$$
$$= \mu_2 + (c-m)^2 > \mu_2.$$

Thus it is seen that the second order moment about the point c will assume its smallest possible value when $c = m$.

If the distribution of the variable X is symmetric about a certain point, the symmetry point must coincide with the centre of gravity of the mass in the distribution, and thus also with the mean value $E(X) = m$. It is easily seen that in this case not only μ_1, but all central moments of odd order, will reduce to zero. This situation occurs in the following examples 1 and 2.

Ex. 1. — If X denotes the number of points obtained in a throw with an ordinary die, we have

$$\alpha_\nu = \frac{1}{6}(1^\nu + 2^\nu + \cdots + 6^\nu).$$

In particular we have $m = \alpha_1 = \frac{7}{2}$, and thus 5.6.3 and 5.6.4 give $\mu_2 = \frac{35}{12}$, $\mu_3 = 0$,

$$\mu_4 = \frac{707}{48}.$$

Ex. 2. — For a rectangular distribution in (a, b), we have

$$\alpha_\nu = \frac{1}{b-a}\int_a^b x^\nu \, dx = \frac{1}{\nu+1} \cdot \frac{b^{\nu+1} - a^{\nu+1}}{b-a}.$$

From 5.6.3 and 5.6.4 we then obtain

$$\mu_2 = \frac{1}{12}(b-a)^2, \quad \mu_3 = 0, \quad \mu_4 = \frac{1}{80}(b-a)^4.$$

Ex. 3. — Find the conditions which a distribution must satisfy in order to have $\mu_2 = 0$ (cf. 6.1).

5.7. Location, dispersion, skewness. — When we are dealing with distributions, it will often be desirable to be able to describe a given distribution by means of a small number of typical values or *characteristics*, which express some essential features of the distribution. Important examples of essential properties, which we shall often want to describe in this way, are the *location* of the distribution, which is usually characterized by means of the abscissa of some appropriately chosen central point for the mass of the distribution, the degree of *dispersion* of the mass about this central value, and finally the degree of asymmetry or *skewness* of the distribution.

The most important *measure of location* is the *mean value* of the distribution, which is identical with the abscissa of the centre of gravity of the mass in the distribution. However, we have seen that there are distributions for which this value does not exist. Moreover, in strongly asymmetrical distributions the position of the centre of gravity is largely determined by the effect of very small quantities of mass situated far out on the axis, so that the mean value does not provide a satisfactory measure for the location of the great bulk of the mass. Distributions of this type sometimes occur in practical applications, e.g. when we are concerned with variables such as income, salaries, capital values of estates, etc.

In such cases it is usually better to use the location measure provided by the *median*, which can be somewhat unprecisely defined as the abscissa of that point of the axis, which divides the mass in the distribution into two equal parts. More precisely, the median is defined as the abscissa of any common point of the line $y = \frac{1}{2}$ and the distribution curve $y = F(x)$, regarding possibly occurring vertical "steps" as parts of the curve. From this definition, it obviously follows that *every distribution has at least one median*. In certain cases the median is uniquely determined; in other cases any point in a certain interval will be a median, and it is then customary to consider the central point of this interval as the proper median of the distribution. The two cases are illustrated in fig. 8–9. Both diagrams show distributions of the discrete type, but the same considerations apply to perfectly general distributions.

For a *symmetric* distribution, such that the mean value exists, it is immediately seen that the mean and the median will be equal.

A further measure of location, which is sometimes useful, is the *mode*. For a continuous distribution this is defined as the abscissa of the maximum point of the frequency curve, while for a discrete distribution the mode coincides with that possible value of the variable which has the largest probability. If there is a unique maximum of the frequence curve

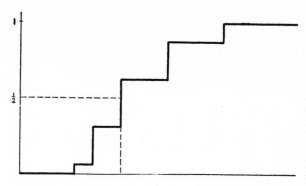

Fig. 8. Distribution function with a uniquely determined median.

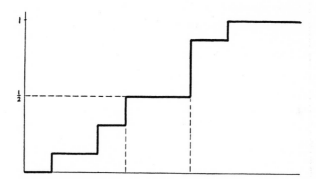

Fig. 9. Distribution function with infinitely many medians. Each x value between the dotted vertical lines is a median.

or the probability values, the mode will be uniquely determined, and in the applications this situation very often occurs. For a symmetric distribution with a single maximum, the mode coincides with the median.

There is one important drawback to the use of location measures such as the median and the mode: for these measures there do not exist any simple rules of calculation, corresponding to the rules of addition and multiplication for mean values given in 5.5.

We now proceed to discuss the *measure of dispersion*. While the mean value m corresponds to the centre of gravity of the mass in the distribution, the central moment of the second order, $\mu_2 = E((X-m)^2)$, corresponds to the *moment of inertia* of the mass with respect to an axis through the centre of gravity, orthogonal to the axis of X. In a case where we use the mean value as our measure of location it would thus seem natural to choose μ_2 as measure of the dispersion of the mass about the mean. However, μ_2 is a quantity of the second order in units of the variable,

77

and accordingly the square root of μ_2 will generally be a more convenient measure of dispersion than μ_2 itself. For an arbitrary random variable X, such that μ_2 exists, this square root will be called the *standard deviation* of the variable or the distribution, and denoted by $D(X)$, or by the single letter σ. Thus we write

$$(5.7.1) \qquad D(X) = \sqrt{\mu_2} = \sigma = \sqrt{E((X-m)^2)},$$

where the square roots should be taken with the positive sign. The square of the standard deviation

$$(5.7.2) \qquad D^2(X) = \mu_2 = \sigma^2 = E((X-m)^2) = E(X^2) - E^2(X)$$

is known as the *variance* of X.

If a and b are constant, we find by 5.5.5

$$D^2(aX+b) = E((aX+b-am-b)^2)$$
$$= a^2 E((X-m)^2) = a^2 D^2(X),$$

and hence

$$(5.7.3) \qquad D(aX+b) = |a| \, D(X).$$

Taking here in particular $a = 1/\sigma$, $b = -m/\sigma$, we obtain from 5.5.5 and 5.7.3

$$E\left(\frac{X-m}{\sigma}\right) = 0, \quad D\left(\frac{X-m}{\sigma}\right) = 1.$$

The variable $(X-m)/\sigma$, which expresses the deviation of X from its mean value, measured in units of the standard deviation, will be called the *normalized* or *standardized* variable corresponding to X.

Further, let X and Y be two *independent variables*. Then

$$D^2(X+Y) = E((X+Y)^2) - E^2(X+Y)$$
$$= E(X^2) - E^2(X) + E(Y^2) - E^2(Y) +$$
$$+ 2E(XY) - 2E(X)E(Y).$$

But according to 5.5.8 we have $E(XY) = E(X)E(Y)$, and consequently

$$(5.7.4) \qquad D^2(X+Y) = D^2(X) + D^2(Y).$$

This important addition rule for standard deviations is immediately extended to an arbitrary number of terms. If $X_1, \ldots, X_n$ are independent variables, we thus have

(5.7.5) $$D^2(X_1 + \cdots + X_n) = D^2(X_1) + \cdots + D^2(X_n).$$

We shall now show an important application of the properties of mean values and standard deviations so far obtained. Let $X_1, \ldots, X_n$ be independent variables, all having *the same mean value m and the same standard deviation* σ. If we form the arithmetic mean $(X_1 + \cdots + X_n)/n$, which we shall denote by $\overline{X}$, we have according to 5.5.5 and 5.5.7

(5.7.6) $$E(\overline{X}) = E\left(\frac{X_1 + \cdots + X_n}{n}\right) = m,$$

and further by 5.7.3 and 5.7.5

(5.7.7) $$D(\overline{X}) = D\left(\frac{X_1 + \cdots + X_n}{n}\right) = \frac{\sigma}{\sqrt{n}}.$$

Thus the arithmetic mean has the same mean value as each of the variables, while the standard deviation has been divided by $\sqrt{n}$.

For distributions of the types mentioned above, where the mean is inconvenient as measure of location, it will practically always, and for the same reasons, be inconvenient to use the standard deviation as measure of dispersion. We can then often use a measure obtained in the following way. In the same way as we defined the median as the point dividing the mass of the distribution into two equal parts, we now define the *first, second* and *third quartiles*, ζ_1, ζ_2 and ζ_3 as the three points that divide the mass in the distribution into four equal parts. This definition can be made entirely precise in the same way as in the case of the median. We thus have the mass $\frac{1}{4}$ to the left of the first quartile, $\frac{1}{4}$ between the first and the second quartiles etc. Obviously the second quartile coincides with the median. The quantity $\zeta_3 - \zeta_1$ is known as the *interquartile range*, and is sometimes used as a measure of dispersion.

A further measure of dispersion, which is getting more and more out of use, but is still found in the literature of certain fields of application, is the *average deviation*. This is the mean value of the absolute deviation of the variable from some central value, as which sometimes the mean and sometimes the median is used. Thus the expression of the average deviation will be

$$E(|X - m|) \quad \text{or} \quad E(|X - \zeta_2|).$$

For this measure there does not exist any simple addition rule of the type 5.7.5. In any case where it is possible to use the standard deviation

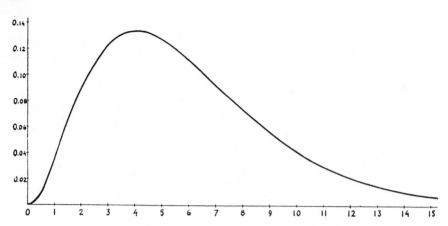

Fig. 10. Distribution with positive skewness. Frequency curve of the χ^2 distribution (cf. 8.1) with $n = 6$ degrees of freedom.

as measure of dispersion it will thus be preferable to do so. And in those cases where the standard deviation is inconvenient as measure of dispersion, the average deviation will usually, and for the same reasons, be inconvenient, so that one has to fall back on the interquartile range.

With respect to the asymmetry or *skewness* of the distribution, we have already found in 5.6 that for a completely symmetric distribution all central moments of odd order will vanish. As a measure of the degree of skewness it will thus seem reasonable to choose one of these central moments, say the central moment of third order μ_3. This is a quantity of the third order in units of the variable, and it is customary to reduce this to a dimensionless constant by dividing by the third power of the standard deviation. The measure of skewness thus obtained will be

$$(5.7.8) \qquad \gamma_1 = \frac{\mu_3}{\sigma^3}.$$

In fig. 10 we show an example of an asymmetric distribution, where the frequency curve has a long "tail" to the positive side. When computing the third moment μ_3 for a distribution of this type, the cubes of the large positive deviations from the mean m will generally outweigh the cubes of the negative deviations, so that the measure of skewness γ_1 takes a positive value. We shall then say that we are concerned with a distribution of *positive skewness*, and vice versa.[1]

---

[1] The terminology on this point is not yet entirely fixed, and some authors define the measure of skewness with the opposite sign.

Finally, the "peakedness" or "excess" of a frequency curve in the central part of the distribution is sometimes measured by the quantity

$$(5.7.9) \qquad\qquad \gamma_2 = \frac{\mu_4}{\sigma^4} - 3.$$

We shall return to this question in connection with the discussion of the normal distribution in ch. 7.

Ex. — We have seen in 5.5, ex. 3, that for the CAUCHY distribution the first moment does not exist. A fortiori the same thing holds for any moment of an order exceeding one. Thus neither the mean, the standard deviation, or the average deviation is available for this distribution. The distribution function is

$$F(x) = \frac{1}{\pi} \int_{-\infty}^{x} \frac{du}{1 + u^2} = \frac{1}{2} + \frac{1}{\pi} \text{ arc tg } x.$$

Thus we have $F(-1) = \frac{1}{4}$, $F(0) = \frac{1}{2}$ and $F(+1) = \frac{3}{4}$, so that the median is equal to zero, while the first and third quartile are respectively -1 and $+1$ (see fig. 6, p. 63). That the median is equal to zero is, of course, an obvious consequence of the fact that the distribution is symmetric about the origin.

5.8. Tchebycheff's theorem. — Let X be a random variable, such that the mean value and the standard deviation of X both exist. The following theorem, which is named after the Russian mathematician TCHE-BYCHEFF, gives a good idea of the significance of the standard deviation as a measure of the dispersion of the distribution.

Denote the mean value of the variable X by m, and the standard deviation by σ. Let further k denote an arbitrary positive number. The probability that the variable X will assume a value satisfying the inequality $|X - m| > k\sigma$ is then smaller than $1/k^2$.

If we have $k < 1$, this is obviously true, since *any* probability will then be smaller than $1/k^2$. On the other hand, if we have $k \geq 1$, the theorem provides an important information about the rate of decrease of the mass in the distribution at large distances from the mean value m. In fact, the theorem asserts that the total amount of mass situated outside the limits $m \pm k\sigma$ is smaller than $1/k^2$, so that in particular this mass will be very small for large values of k.

The theorem holds true for a perfectly arbitrary distribution, but we shall here only give the proof for a distribution of the continuous type, leaving to the reader to work out a corresponding proof for the discrete type.

Wen x falls outside the limits $m \pm k \sigma$, we have everywhere $(x-m)^2 > k^2 \sigma^2$; within the same limits we have everywhere $(x-m)^2 > 0$, except in the point $x = m$, where the sign of equality holds in the last relation. Hence we obtain

$$\sigma^2 = \int_{-\infty}^{\infty} (x-m)^2 f(x)\, dx > k^2 \sigma^2 \left(\int_{-\infty}^{m-k\sigma} f(x)\, dx + \int_{m+k\sigma}^{\infty} f(x)\, dx \right).$$

However, the sum of the two integrals within the brackets in the last member is identical with $P(|X-m| > k\sigma)$, the probability that the variable X assumes a value such that its absolute deviation from the mean value m exceeds $k\sigma$. Thus we have the desired inequality

$$P(|X-m| > k\sigma) < \frac{1}{k^2}.$$

More generally, let us write

$$\lambda^2 = E(X-c)^2,$$

where c is any given constant (thus in particular for $c = m$, we have $\lambda = \sigma$). It is then easily shown in the same way as above that

$$P(|X-c| > k\lambda) < \frac{1}{k^2}.$$

If λ is small, it follows that the bulk of the mass in the X distribution is situated in a small interval containing the point c.

5.9. Addition of independent variables. — It is often important to consider the distribution of the sum

$$X = X_1 + X_2 + \cdots + X_n$$

of a certain number of independent variables $X_1, X_2, \ldots, X_n$ having known distributions. If the mean values and the standard deviations of the variables X and $X_1, \ldots, X_n$ are denoted respectively by $m, m_1, \ldots, m_n$ and $\sigma, \sigma_1, \ldots, \sigma_n$, we have according to 5.5.7 and 5.7.5

(5.9.1)
$$m = m_1 + \cdots + m_n,$$
$$\sigma^2 = \sigma_1^2 + \cdots + \sigma_n^2.$$

In order to investigate the distribution of the sum X a little more closely, we first consider the case $n = 2$, where we have

$$X = X_1 + X_2,$$

X_1 and X_2 being independent. Then the joint distribution of X_1 and X_2 defines a certain distribution of mass in the (X_1, X_2)-plane, and the probability that X assumes a value $\leq x$ is equal to the total amount of mass situated in the domain $X_1 + X_2 \leq x$. However, this probability is identical with the distribution function of X, so that these remarks provide a method for deducing an expression of this function in the general case.

As in the preceding section we shall here only consider the continuous case, leaving the discrete case to the reader. If the distribution functions of X, X_1 and X_2 are respectively F, F_1 and F_2, while the corresponding frequency functions are f, f_1 and f_2, the probability density of the joint distribution of X_1 and X_2 will, according to 5.4.9, be $f_1(x_1) f_2(x_2)$. From the above remarks it then follows that we have

(5.9.2)
$$\begin{aligned}
F(x) &= \iint\limits_{u+v \leq x} f_1(u) f_2(v)\, du\, dv \\
&= \int_{-\infty}^{\infty} f_2(v)\, dv \int_{\infty}^{x-v} f_1(u)\, du.
\end{aligned}$$

If the frequency function $f_1(x)$ is bounded for all values of x, it is allowed to differentiate under the integral sign, and we obtain the important relation

(5.9.3)
$$\begin{aligned}
f(x) = F'(x) &= \int_{-\infty}^{\infty} f_1(x-v) f_2(v)\, dv \\
&= \int_{-\infty}^{\infty} f_2(x-u) f_1(u)\, du.
\end{aligned}$$

By a repeated application of this formula, we can find the frequency function of the sum of three, four, or any number of independent variables, all having known frequency functions.

Ex. 1. — If X_1 and X_2 are independent, and both have rectangular distributions in $(0, 1)$, it is easily seen that the probability density of the joint distribution will be constantly equal to 1 within the square $0 < X_1 < 1$, $0 < X_2 < 1$, and equal to 0 outside this square (cf. 5.4, ex. 2). The amount of mass within the domain $X_1 + X_2 \leq x$ will then be equal to the area of that part of the square which belongs to this domain. It follows that the distribution function $F(x)$ of the sum $X_1 + X_2$ is equal to 0 for $x < 0$, and equal to 1 for $x > 2$, while for $0 < x < 2$ we have

$$F(x) = \begin{cases} \frac{1}{2} x^2 & \text{for } 0 < x < 1, \\ 1 - \frac{1}{2}(2-x)^2 & \text{for } 1 < x < 2. \end{cases}$$

By differentiating it is found that the corresponding frequency function $f(x)$ is 0 for $x < 0$ and for $x > 2$, while

$$f(x) = \begin{cases} x & \text{for } 0 < x < 1, \\ 2 - x & \text{for } 1 < x < 2. \end{cases}$$

The corresponding frequency curve has the shape of a "roof", or a triangle over the base $0 < X < 2$, and the distribution is known as a *triangular distribution*.

Find the mean value and the standard deviation of this distribution, using 5.5.6 and 5.7.4!

Study the distribution of the sum of three or four independent variables, all having rectangular distributions over (0, 1)!

What becomes of the above distributions when the interval (0, 1) is replaced by an arbitrary interval (a, b)? (Use here 5.3.4 and 5.3.5).

Ex. 2. — If X_1 and X_2 both have CAUCHY distributions, the frequency function of $X_1 + X_2$ is by 5.9.2

$$f(x) = \frac{1}{\pi^2} \int\limits_{-\infty}^{\infty} \frac{du}{[1 + (x - u)^2] (1 + u^2)}.$$

The integral can be calculated by means of standard methods for the integration of rational functions (the reader should work out this calculation!), and we find

$$f(x) = \frac{2}{\pi (x^2 + 4)}.$$

By 5.3.5, the arithmetic mean $\frac{1}{2}(X_1 + X_2)$ will then have the frequency function

$$2f(2x) = \frac{1}{\pi (1 + x^2)},$$

which is identical with the frequency function of each of the original variables X_1 and X_2. By a repetition of this procedure it is found that the same property holds true for sums of more than two variables. If $X_1, \ldots, X_n$ are independent variables all having CAUCHY distributions, the arithmetic mean $\frac{1}{n}(X_1 + \cdots + X_n)$ thus has the same distribution as each of the variables X_i. This remarkable property of the CAUCHY distribution is essentially connected with the fact that the first and second order moments of this distribution do not exist (cf. 7.3).

5.10. Generating functions. — We define the *generating function* $\psi(t)$ for a random variable X, or for the corresponding distribution, by writing

(5.10.1) $$\psi(t) = E(e^{tX}).$$

where t is a real auxiliary variable.[1] For a distribution of the discrete type we have

$$(5.10.2) \qquad \psi(t) = \sum_{\nu} p_{\nu} e^{t \, x_{\nu}},$$

and for a distribution of the continuous type

$$(5.10.3) \qquad \psi(t) = \int_{-\infty}^{\infty} e^{t \, x} f(x) \, dx.$$

For $t = 0$ the sum and the integral are always convergent, and we have by 5.2.1 and 5.3.3

$$\psi(0) = 1.$$

On the other hand, when $t \neq 0$, neither the sum nor the integral will be necessarily convergent. Nevertheless, we actually find convergence, at least for some values of t, in the case of many important distributions. If, e.g., there exists a finite quantity A such that the variable X can only assume values exceeding $-A$, it is easily seen that the series and the integral will both converge for $t \leq 0$. (In this case, it is only necessary to extend the integral from $-A$ to $+\infty$). Even if no such finite value A exists, the same conclusion will hold, e.g., in the continuous case, as soon as $f(x)$ tends sufficiently strongly to zero as $x \to -\infty$.

The convergence difficulties can be entirely avoided, if we allow the variable t to assume complex values. If, in the expression 5.10.1 of the generating function, t is replaced by it, we obtain the function

$$\varphi(t) = \psi(it) = E(e^{it \, X}),$$

which is known as the *characteristic function* of the variable X. For distributions of the two simple types $\varphi(t)$ is represented by a series or an integral, corresponding to 5.10.2 or 5.10.3, and these will always converge for all real values of t. Even for a perfectly general distribution, it is possible to obtain an analogous expression which is always convergent. On account of these properties the characteristic functions have important applications in the general theory of distributions. However, in this book we do not want to introduce complex variables, and accordingly we shall only deal with the generating function $\psi(t)$. We now proceed to deduce some properties of this function, which hold true for all distributions such that the series 5.10.2, or the integral 5.10.3, converges for at least some value of $t \neq 0$.

---

[1] Sometimes the generating function is a little differently defined, viz. as the mean value $E(t^X)$ obtained if, in 5.10.1, we replace e^t by t.

If the moment α_ν of order ν exists, we obtain by repeated differentiation of 5.10.2 or 5.10.3, taking after the differentiation $t = 0$,

(5.10.4) $$\psi^{(\nu)}(0) = \alpha_\nu.$$

Thus if all moments α_ν exist, the MACLAURIN series of $\psi(t)$ vill be

(5.10.5) $$\psi(t) = \sum_0^\infty \frac{\alpha_\nu}{\nu!} t^\nu.$$

If this series is convergent for some $t \neq 0$, it can be shown that be generating function $\psi(t)$, or the sequence of moments α_ν, determines the corresponding distribution uniquely. For the proof of this important fact we must refer, e.g., to *Mathematical Methods*, Ch. 15.

The generating function of a linear function $Y = aX + b$ is

(5.10.6) $$E(e^{t(aX+b)}) = e^{bt} E(e^{atX}) = e^{bt} \psi(at).$$

Let now X_1 and X_2 be independent variables, and consider the sum $X = X_1 + X_2$. If the generating functions of X, X_1 and X_2 are ψ, ψ_1 and ψ_2, we have according to 5.5.8

$$E(e^{tX}) = E(e^{t(X_1+X_2)}) = E(e^{tX_1}) \cdot E(e^{tX_2}),$$

or

(5.10.7) $$\psi(t) = \psi_1(t)\,\psi_2(t).$$

This relation is directly extended to sums of more than two variables. *Thus the generating function of a sum of independent variables is the product of the generating functions of the terms.*

Ex. — An urn contains k balls, numbered from 1 to k. From this urn, n balls are drawn with replacement. What is the probability that the sum of the numbers obtained in the course of these drawings will assume a given value s?

For the number X_ν obtained in the ν:th draving, we have the possible values 1, 2, ..., k, each having the probability $1/k$. The generating function of X_ν will thus be $(e^t + \cdots + e^{kt})/k$, and according to the above proposition the generating function of the sum $X = X_1 + \cdots + X_n$ will be

$$\left(\frac{e^t + e^{2t} + \cdots + e^{kt}}{k}\right)^n = \sum_{r=n}^{kn} p_r\, e^{rt}.$$

The required probability is thus identical with the coefficient p_s of e^{st} in this development.

5.11. Problems. — **1.** A random variable X has the d.f. $F(x)$. a) Find the d.f:s of the variables $Y = X^2$ and $Z = e^X$. b) Assuming that the X distribution belongs to the continuous type, find the fr.f:s of Y and Z.

2. A number X is randomly chosen[1] between 0 and 1. Find the d.f:s and fr.f:s of the variables X^2, X and e^X, and draw diagrams of the corresponding distribution curves and frequency curves.

3. O is the origin of an orthogonal coordinate system, P is a variable point on the x axis, and Q is the point $x = 0$, $y = 1$. Suppose that the angle OQP is randomly chosen between $-\frac{1}{2}\pi$ and $+\frac{1}{2}\pi$. Find the d.f. and the fr.f. of the abscissa of P.

4. (a, b) is a given interval. In this interval, n numbers are randomly chosen and arranged by increasing order of magnitude. Denote these numbers by $X_1 < X_2 < \cdots < X_n$, and find the d.f., the fr.f. and the mean value of any X_i.

5. Find the modes of the distributions of PEARSON's types I, III and VI (cf. 8.4).

6. X is a variable with an arbitrary distribution. What value should be given to the constant c in order to render the mean value $E|X - c|$ as small as possible?

7. At every performance of the random experiment E, the probability of the occurrence of the event A is equal to p. E is repeated until A occurs for the first time. Find the mean value and the standard deviation of the number n of repetitions required.

8. An urn contains N balls, Np of which are white, while the other $Nq = N(1-p)$ are black. n balls are drawn without replacement, and among these there are f white and $n - f$ black balls. Find the mean and the s.d. of f, and compare the results with the corresponding values for drawings with replacement, which are given by 6.2.3. Hint: write $f = f_1 + \cdots + f_n$, where f_i denotes the number of white balls in the i:th drawing (thus $f_i = 0$ or 1), and find $E(f_i)$, $E(f_i^2)$ and $E(f_i f_k)$. (cf. also 16.5.)

9. The total number of points obtained in n throws with an ordinary die is denoted by X. Find $E(X)$ and $D(X)$, and form the standardized variable corresponding to X.

10. Find the mean and the s.d. of the number of rencontres in the game of rencontre discussed in 4.3.

11. X and Y are randomly chosen between 0 and 1. Find the d.f., the fr.f. and the mean value of a) XY, b) $X - Y$, c) $|X - Y|$.

12. a, b and c are randomly chosen between 0 and 1. Find the probability that the equation $ax^2 + 2bx + c = 0$ will have real roots.

13. Same question for the equation $ax^2 + bx + c = 0$.

14. A point P is randomly chosen on the perimeter of a circle with radius r. Find the mean value of the distance AP between P and a given point A on the perimeter.

15. P and Q are randomly chosen as in the preceding exercise. Find the mean value of the distance PQ.

16. P, Q and R are randomly chosen as in the preceding exercise. a) Find the probability that at least one of the angles of the triangle PQR will exceed $k\pi$, where k is

---

[1] For the sake of brevity, this expression will be used henceforth to denote that X is assumed to be uniformly distributed (cf. 5.3, ex. 1) over the interval $(0, 1)$. We shall use corresponding expressions when a point X is taken on an arbitrary interval, on a circumference, etc. When two or more points or numbers are chosen in this way, it will always be assumed that the corresponding variables are independent, unless explicitly stated otherwise.

a given constant such that $0 < k < 1$. b) Find the fr.f. for the greatest angle of the triangle.

17. Show that for any two random variables X and Y we have $[E(XY)]^2 \leq E(X^2)E(Y^2)$. (SCHWARZ's inequality.) Hence deduce in particular that $\mu_2^2 \leq \mu_4$ for any distribution.

18. X is a non-negative random variable with the mean m. Show that

$$P(X > km) < 1/k$$

for any $k > 0$.

19. Same assumptions as in problem 7 above. The experiment E is repeated until an uninterrupted run of ν occurrences of A has been observed. Let the number of repetitions required be denoted by n, and find the generating function and the mean value of n.

20. Same assumptions as in 4.4, problem 21. A person goes on buying packages of cigarrettes until he gets a complete set of all k numbers. Find the generating function and the mean value of the number of packages required.

21. $X_1, \ldots, X_m$ and $Y_1, \ldots, Y_n$ are $m+n$ independent random variables, while $a_1, \ldots, a_m$ and $b_1, \ldots, b_n$ are constants. Then the variables

$$X = a_1 X_1 + \cdots + a_m X_m \quad \text{and} \quad Y = b_1 Y_1 + \cdots + b_n Y_n$$

are independent. (This statement is true for all kinds of distributions. The reader should work out a proof for the particular case when all the X_i and Y_i have distribubutions of the continuous type.)

22. $X_1, \ldots, X_n$ are independent random variables, none of which has zero s.d. Let $a_1, \ldots, a_n$ and $b_1, \ldots, b_n$ be non-negative constants, and write

$$U = a_1 X_1 + \cdots + a_n X_n, \quad V = b_1 X_1 + \cdots + b_n X_n.$$

In order that U and V may be independent, it is necessary and sufficient that $a_i b_i = 0$ for all $i = 1, 2, \ldots, n$.

APPLICATIONS TO MORTALITY TABLES AND LIFE INSURANCE

23. The probability that a person aged x years will die before attaining the age $x + dx$ may reasonably be assumed to tend to zero with dx, and to be approximately proportional to dx when the latter is small. For an infinitely small dx, this probability may accordingly be denoted by $\mu_x dx$, where μ_x is a factor depending on x. The quantity μ_x is known as the *force of mortality* at age x. Show that, in the notation of 3.7, ex. 15,

$$\mu_x = -\frac{1}{l_x} \frac{dl_x}{dx},$$

$$_t p_x = \frac{l_{x+t}}{l_x} = e^{-\int_0^t \mu_{x+u}\, du}$$

24. Consider a group of l_x persons of age x, who may be considered mutually independent in respect of their chances of life and death. Let λ_{x+t} denote the number of those persons in the group who will survive at age $x + t$. Show that

$$E\left(\lambda_{x+t}\right)=l_{x+t}.$$

(We anticipate here formula 6.2.3 of the following chapter.)

25. Let $x+y$ denote the age at death of a person now aged x. Find the fr.f. of y, and show that

$$E\left(y\right)=\frac{1}{l_x}\int_0^\infty l_{x+t}\,dt.$$

This mean value is known as the *expectation of life* of a person aged x.

26. The present value of a sum s payable after t years is $s\,e^{-\delta t}$, where δ is the *force of interest*. Suppose that a person now aged x is entitled to receive the sum 1 after t years, on the condition that he is then alive. Let z be the present value of the payment due to him. Show that

$$E\left(z\right)=\frac{l_{x+t}}{l_x}\,e^{-\delta t}=\frac{D_{x+t}}{D_x},$$

where

$$D_x=l_x\,e^{-\delta x}.$$

27. A person aged x is entitled to receive a *temporary life annuity* payable continuously during at most n years, the amount dt falling due during the time element dt, each payment being subject to the condition that he is alive at the moment the payment is made. If u is the total present value of the payments thus due to him, the mean value $E\left(u\right)$ is usually denoted by $\bar{a}_{x\overline{n}|}$. Show that

$$\bar{a}_{x\overline{n}|}=\int_0^n\frac{D_{x+t}}{D_x}\,dt=\frac{\overline{N}_x-\overline{N}_{x+n}}{D_x},$$

where

$$\overline{N}_x=\int_x^\infty D_t\,dt.$$

28. A person aged x contracts an *endowment insurance*, by which he is entitled to receive the sum 1 on attaining the age $x+n$, or to have the same sum paid to his heirs at the moment of his death, if this occurs before the age $x+n$. If v is the present value of the payment due to him under this contract, the mean value $E\left(v\right)$ is usually denoted by $\overline{A}_{x\overline{n}|}$. Show that

$$\overline{A}_{x\overline{n}|}=1-\delta\,\bar{a}_{x\overline{n}|}.$$

29. In return of the insurance contract discussed in the preceding exercise, the insured person agrees to pay the *premium* $p\,dt$ during each time element dt, until the insurance sum will fall due. This may be regarded as a game of chance between the insured person, who undertakes to pay the premiums, and the contracting insurance company. Show that, if we take

$$p=\frac{\overline{A}_{x\overline{n}|}}{\bar{a}_{x\overline{n}|}}=\frac{1}{\bar{a}_{x\overline{n}|}}-\delta,$$

the expectation of each party will be zero, so that the game is a "fair" one (cf. 6.3).

CHAPTER 6

THE BINOMIAL AND RELATED DISTRIBUTIONS

6.1. One and two point distributions. — The simplest conceivable discrete distribution is one where the total mass is concentrated in one single point m. A random variable X having a distribution of this type is not, properly speaking, a "variable" at all, since we have the probability 1 for the event that the variable assumes the value m, and the probability 0 that X assumes any other value. Obviously this distribution has the mean value m and the s.d. 0, and it is easily seen that this is the only possible distribution having a s.d. equal to zero. If we define a function $\varepsilon(x)$ by writing

$$(6.1.1) \qquad \varepsilon(x) = \begin{cases} 0 & \text{for } x < 0, \\ 1 & \text{for } x \geq 0, \end{cases}$$

$\varepsilon(x)$ will be the d.f. of a distribution having the total mass concentrated at the origin, and $\varepsilon(x - m)$ will then correspond to the distribution where the total mass is situated in the point m.

Consider now the case of a distribution, tho total mass of which is concentrated in two different points. By an appropriate choice of origin and scale for the variable X, these points can always be made to correspond to the values $X = 0$ and $X = 1$. Let the amount of mass in the point $X = 1$ be denoted by p, while the point $X = 0$ carries the mass $q = 1 - p$. For a variable X having this distribution, the only possible values are 1 and 0, the corresponding probabilities being p and q respectively. A variable of this kind is obtained, e.g., if we define X as the number of occurrences of an event A of probability p in one single performance of the random experiment to which A is attached. The mean and the variance of X are

$$(6.1.2) \qquad \begin{aligned} E(X) &= p \cdot 1 + q \cdot 0 = p, \\ D^2(X) &= p(1 - p)^2 + q(0 - p)^2 = pq. \end{aligned}$$

The generating function is

$$(6.1.3) \qquad \psi(t) = p\, e^t + q.$$

6.2. The binomial distribution. — Let f denote the absolute frequency of the event A in a series of n independent repetitions of a certain random experiment, where in each repetition we have the probability $p = 1 - q$ that A occurs. The frequency f is then a random variable with the possible values $0, 1, \ldots, n$, and we have seen in 4.1 that the probability that f will assume any given possible value ν is expressed by the binomial formula

$$(6.2.1) \qquad p_\nu = \binom{n}{\nu} p^\nu q^{n-\nu}.$$

The same expression can be simply deduced by means of generating functions. Let f_ν denote the number of times that A occurs *in the ν:th repetition*. Then f_ν has the only possible values 1 and 0, and by 6.1.3 the generating function of f_ν is $p e^t + q$. Further, $f = f_1 + \cdots + f_n$, where all f_ν are independent, and hence by 5.10 the generating function of the sum f is

$$(6.2.2) \qquad \sum_0^n p_\nu e^{\nu t} = (p e^t + q)^n = \sum_1^n \binom{n}{\nu} p^\nu q^{n-\nu} e^{\nu t},$$

so that we are again led to the expression 6.2.1.

For given values of p and n, the random variable f will thus have a distribution of the discrete type, such that the point ν carries the mass p_ν given by 6.2.1, for $\nu = 0, 1, \ldots, n$. This distribution is known as the *binomial distribution*.

The ratio between two consecutive values of p_ν,

$$\frac{p_{\nu+1}}{p_\nu} = \frac{n-\nu}{\nu+1} \cdot \frac{p}{q},$$

is $\gtrless 1$, according as $\nu \lessgtr np - q$. Suppose first that $np - q$ is not an integer, and let r denote the uniquely determined integer which is then situated between the limits $np - q$ and $np - q + 1 = np + p$. The probabilities p_ν are then steadily increasing as ν increases from 0 to r, and steadily decreasing as ν increases from r to n. The ordinates in the probability diagram of the distribution will thus show a single maximum for $\nu = r$, which corresponds to the mode of the distribution. — On the other hand, if we suppose that $np - q$ is an integer, it is readily seen that there will be two equal maximum ordinates corresponding to $\nu = np - q$ and $\nu = np + p$. In figs. 11–12, one diagram is shown for each case.

According to 6.1.2, each of the variables f_ν introduced above has the mean p and the standard deviation $\sqrt{pq}$. For the sum $f = f_1 + \cdots + f_n$, we then obtain from 5.5.7 and 5.7.5

$$(6.2.3) \qquad E(f) = np. \qquad D(f) = \sqrt{npq}.$$

91

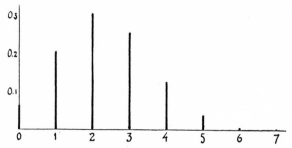

Fig. 11. Binomial distribution with a single maximum term. $p=\frac{1}{3}$, $n=7$.

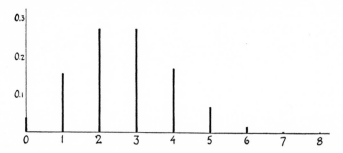

Fig. 12. Binomial distribution with two equal maximum terms. $p=\frac{1}{3}$, $n=8$.

Thus the mean of the binomial distribution is np, while the standard deviation is $\sqrt{npq}$. In order to find the higher central moments $\mu_\nu = E((f-np)^\nu)$, we form according to 5.10.6 the generating function of the difference $f-np$, which is

(6.2.4)
$$\psi(t) = e^{-npt}(pe^t + q)^n = (pe^{qt} + qe^{-pt})^n$$
$$= \left(\sum_0^\infty \frac{pq^\nu + q(-p)^\nu}{\nu!}t^\nu\right)^n.$$

However, since the central moments μ_ν are identical with the moments of $f-np$, we obtain from 5.10.5

$$\psi(t) = \sum_0^\infty \frac{\mu_\nu}{\nu!}t^\nu.$$

so that we can find an expression of μ_ν by equating the coefficients of t^ν in the two expansions of $\psi(t)$. In this way we obtain

(6.2.5)
$$\mu_2 = npq,$$
$$\mu_3 = npq(q-p),$$
$$\mu_4 = 3n^2p^2q^2 + npq(1-6pq).$$

92

The measures of skewness and excess are, by 5.7.8 and 5.7.9,

$$(6.2.6) \qquad \gamma_1 = \frac{q-p}{\sqrt{npq}}, \qquad \gamma_2 = \frac{1-6pq}{npq}.$$

It will be seen that both these measures tend to zero as $n \to \infty$. The skewness is positive for $p < \frac{1}{2}$, negative for $p > \frac{1}{2}$.

6.3. Bernoulli's theorem. — Using the same notation as in the preceding section, we now consider the frequency ratio f/n of the event A in our series of n independent observations. The mean and the standard deviation of f/n are obtained from 6.2.3, using 5.5.5 and 5.7.3:

$$(6.3.1) \qquad E\left(\frac{f}{n}\right) = p, \qquad D\left(\frac{f}{n}\right) = \sqrt{\frac{pq}{n}}.$$

Let now ε denote a given, arbitrarily small positive quantity. We apply TCHEBYCHEFF's theorem (5.8) to the random variable f/n, taking $k = \varepsilon \sqrt{\dfrac{n}{pq}}$. We then obtain, since pq is always $\leqq \frac{1}{4}$,

$$P\left(\left|\frac{f}{n} - p\right| > \varepsilon\right) < \frac{pq}{n\varepsilon^2} \leqq \frac{1}{4n\varepsilon^2}.$$

Thus for any given positive ε the probability that the absolute deviation of the frequency ratio f/n from the corresponding probability p will exceed ε tends to zero as $n \to \infty$.

This is the famous theorem of BERNOULLI mentioned above in the historical introduction (cf. 1.4).

A more general case of a series of n independent experiments was considered by POISSON, who allowed the probability of the occurrence of the event A to vary from one experiment to another. If the probability of A in the ν:th experiment is $p_\nu = 1 - q_\nu$, and if as before we write $f = f_1 + \cdots + f_n$, where f_ν denotes the number of occurrences of A in the ν:th experiment, we find

$$E(f) = \sum_1^n p_\nu, \qquad D(f) = \sqrt{\sum_1^n p_\nu q_\nu}.$$

Applying TCHEBYCHEFF's theorem to the random variable f/n we then obtain, denoting by $\bar{p} = \dfrac{1}{n}\sum_1^n p_\nu$ the average probability of A in the series of n experiments,

$$P\left(\left|\frac{f}{n} - \overline{p}\right| > \varepsilon\right) < \frac{\sum p_\nu q_\nu}{n^2 \varepsilon^2} \leq \frac{1}{4 n \varepsilon^2}.$$

Thus again the probability that the absolute deviation of the frequency ratio f/n from the average probability $\overline{p}$ exceeds any given positive ε tends to zero, as $n \to \infty$.

A further generalization is obtained in the following way. Let, for every integer n, the variables $X_1, \ldots, X_n$ be independent and such that $E(X_\nu) = m_\nu$ and $D(X_\nu) = \sigma_\nu$. Write

$$\overline{X} = \frac{1}{n} \sum_1^n X_\nu, \qquad \overline{m} = \frac{1}{n} \sum_1^n m_\nu.$$

Then $\overline{X}$ has the mean $\overline{m}$ and the standard deviation $\dfrac{1}{n} \sqrt{\sum_1^n \sigma_\nu^2}$, so that TCHEBYCHEFF's theorem gives

$$P(|\overline{X} - \overline{m}| > \varepsilon) < \frac{1}{n^2 \varepsilon^2} \sum_1^n \sigma_\nu^2.$$

As soon as $\sum_1^n \sigma_\nu^2 / n^2$ tends to zero as $n \to \infty$, the same will thus hold for the probability that the absolute deviation $|\overline{X} - \overline{m}|$ will exceed any given positive ε. — If, in particular, each X_ν has the only possible values 1 and 0, we obtain as a special case the POISSON series of experiments discussed above.

In order to give an application of the last result, we now consider the gain of a player in a series of repetitions of a certain game of chance at a gambling establishment. Suppose that, in each performance of the game, the player is allowed to stake an amount at his own choice, up to a certain maximum K fixed by the bank. According to the outcome of the game, the player then either looses his stake, or wins a certain amount, which in each case is proportional to his stake. The amount of the gain in one single game, per unit of stake, is a random variable X. We write as usual $E(X) = m$ and $D(X) = \sigma$. Suppose now that our man plays n games, with the stakes $a_1, \ldots, a_n$. Let $\overline{a} = \sum a_\nu / n$ denote the average amount at stake in each game, while X_ν denotes the gain in the ν:th game. Then $E(X_\nu) = a_\nu m$ and $D(X_\nu) = a_\nu \sigma$. Further, we obviously have $\overline{a} \leq K$. The variable $\overline{X}$ is the average gain per game, and our formula gives

$$P(|\overline{X} - \overline{a}m| > \varepsilon) < \frac{\sigma^2 \sum a_\nu^2}{n^2 \varepsilon^2} \leq \frac{\sigma^2 K^2}{n \varepsilon^2},$$

or conversely

$$P(|\overline{X} - \overline{a}m| \leq \varepsilon) > 1 - \frac{\sigma^2 K^2}{n \varepsilon^2}.$$

In all games occurring in practice the mean m of the gain of the player is negative, which implies that the mean of the gain of the *bank* is positive. For large values of n, it then follows from the last relation that the player has a probability which only differs insignificantly from 1 to make an *average loss* of approximately $\overline{a}|m|$ per game. And our deduction shows that this result holds true *quite independently of how the stakes are fixed within the prescribed maximum.* A game of this type is *disadvantageous* to the player. On the other hand, if $m > 0$, the game is *advantageous*, and finally for $m = 0$ we talk of a *fair* game. In the last case, it is not possible to draw any definite inference about the sign of the average gain.

If, in particular, we take all $a_\nu = 1$, we find that for large n the average gain $\overline{X}$ will, with a probability almost equal to 1, approximately coincide with the mean m. Accordingly the quantity m is sometimes called the *mathematical expectation* of the player in each game, per unit of amount at stake.

Ex. 1. — The game of *rouge et noir* is played with a roulette, divided into 37 cases of equal size, numbered from 0 to 36. 18 of these cases are red, 18 are black, and one, the zero, is of some other colour. If a player stakes a certain amount on red, and red wins, he gets his stake back, and in addition he receives an equal amount from the bank; in the opposite case he looses his stake. The amount of the stakes are left at the player's choice, within the maximum K fixed by the bank.

Show that, in the notation used above, $E(X) = -\dfrac{1}{37}$ and $D^2(X) = \dfrac{1368}{1369}$. Further, criticise the following argument: "a player continually stakes on red. In the first game, he stakes an amount a. As soon as he looses a game, he doubles his stake in the following game; as soon as he wins a game he again stakes the amount a in the following game. If he stops playing after having won n times, his total gain will be na, so that by going on sufficiently long he can finally earn any amount desired."

Ex. 2. — For the validity of the above remarks concerning advantageous and disadvantageous games it is essential that we consider a long series of repetitions of the same game. If, on the other hand, we only consider one single or a small number of repetitions, it is by no means certain that, e.g., a game with a positive m will be advantageous in the everyday sense of this word. A good example is afforded by the famous *Petersburg paradox*, which about 1730 was discussed by DANIEL BERNOULLI in the transactions of the Petersburg academy. A person is offered to play at heads and tails on the condition that he will receive the amount of 2^n if head appears for the first time in the n:th throw. The mean value of the amount that he will thus receive is $\dfrac{2}{2} + \dfrac{2^2}{2^2} + \dfrac{2^3}{2^3} + \cdots$, i.e., infinitely large. He may thus pay any prize required in order to enter the game, and the game will still be "advantageous", in the sense given above that the mean of his gain is positive, and even infinitely large. On the other hand, the simplest common sense seems to show that it will be foolish to risk even a moderate sum on this game, and this apparent paradox has given rise to a considerable number of more or less complicated explanations.

However, the paradox will appear less evident as soon as we realize the fundamental difference between the chances of the player, according as we only consider one single game, or a series consisting of a large number of repetitions of the same game. Suppose, e.g., that our man pays the amount of 2^N, and then plays *one single game* on the conditions stated above. If there appears at least one head in the first $N-1$ throws, he will then certainly suffer a loss, and the probability of this event is $1 - (\frac{1}{2})^{N-1}$. Now, as soon as N is at least moderately large, a game where the probability of making a loss is $1 - (\frac{1}{2})^{N-1}$ is certainly not advantageous in the everyday sense of this word, in spite of the positive mean value of the gain. On the other hand, if it is agreed from the beginning that the game will be repeated a large number of times, it follows from the above that, for any given N, the probability of realizing a positive gain can be made to lie as close to 1 as we please by taking the number of repetitions sufficiently large. A more detailed analysis shows, however, that for a given and moderately large value of N it is necessary to play a very long series of repetitions in order to have a probability of, say, 99 % to make a positive gain.

6.4. De Moivre's theorem. — BERNOULLI's theorem was first published in 1713, more than 150 years before the theorem of TCHEBYCHEFF, which has been used for the proof given in the preceding section. Originally, the theorem was proved by means of a direct analysis of the explicit expression of the probability that the frequency f is situated between two given limits. Using the same notations as in the preceding section, we have by 4.1.2

$$P\left(\left|\frac{f}{n}-p\right|\leq\varepsilon\right)=P(|f-np|\leq n\varepsilon)$$

(6.4.1)

$$=\sum_{n(p-\varepsilon)\leq\nu\leq n(p+\varepsilon)}\binom{n}{\nu}p^\nu q^{n-\nu}.$$

By a painstaking evaluation of the binomial terms appearing in the last member, BERNOULLI deduced the result that the sum tends to 1, as $n\to\infty$, which is equivalent to the theorem. He tells us himself that it took him twenty years to complete the proof of this result, which shows the degree of difficulty involved. Since the time of BERNOULLI, the methods of proof of asymptotic relations of this type have been largely improved, so that it is now possible to deduce by relatively simple arguments a result considerably stronger than the original BERNOULLI theorem. We shall now deduce this result, which was first given by DE MOIVRE as early as 1733.

It will be practical to introduce the standardized variable (5.7) corresponding to the frequency f, which we denote by

$$\lambda=\frac{f-np}{\sqrt{npq}},$$

assuming $0<p<1$. We then have

$$E(\lambda)=0,\qquad D(\lambda)=1.$$

The probability that λ lies between two given limits λ_1 and λ_2 is then identical with the probability that f lies between the limits $np+\lambda_1\sqrt{npq}$ and $np+\lambda_2\sqrt{npq}$. Consequently we have

$$(6.4.2) \qquad P(\lambda_1 < \lambda \leqq \lambda_2) = \sum \binom{n}{\nu} p^\nu q^{n-\nu},$$

where the last sum should be extended over all values of ν satisfying the conditions

$$(6.4.3) \qquad np + \lambda_1 \sqrt{npq} < \nu \leqq np + \lambda_2 \sqrt{npq}.$$

We now allow the number n of repetitions to increase indefinitely, while p, λ_1 and λ_2 are kept fixed. As n tends to infinity, the difference between the limits for ν imposed by 6.4.3 will then also tends to infinity, and so will the number of terms in the sum in the last member of 6.4.2. *However, we are now going to show that the sum of all these terms, which is identical with the probability* $P(\lambda_1 < \lambda \leqq \lambda_2)$, *tends to a definite limiting value as* $n \to \infty$.

The general term of the sum in 6.4.2 has the expression

$$p_\nu = \binom{n}{\nu} p^\nu q^{n-\nu} = \frac{n!}{\nu!\,(n-\nu)!} p^\nu q^{n-\nu}.$$

If we now use STIRLING's formula (cf. Appendix, 4), we find after some reductions

$$p_\nu = \frac{1}{\sqrt{2\pi npq}} \left(\frac{np}{\nu}\right)^{\nu+\frac{1}{2}} \left(\frac{nq}{n-\nu}\right)^{n-\nu+\frac{1}{2}} e^{\frac{1}{12}\left(\frac{\theta_1}{n} - \frac{\theta_2}{\nu} - \frac{\theta_3}{n-\nu}\right)},$$

where θ_1, θ_2 and θ_3 all lie between 0 and 1. Writing

$$(6.4.4) \qquad \nu = np + \lambda \sqrt{npq},$$

we then have

$$p_\nu = \frac{1}{\sqrt{2\pi npq}} e^{-z},$$

where

$$z = (\nu + \tfrac{1}{2}) \log \frac{\nu}{np} + (n-\nu+\tfrac{1}{2}) \log \frac{n-\nu}{nq} - \frac{1}{12}\left(\frac{\theta_1}{n} - \frac{\theta_2}{\nu} - \frac{\theta_3}{n-\nu}\right)$$

$$= (np + \lambda \sqrt{npq} + \tfrac{1}{2}) \log \left(1 + \frac{\lambda q}{\sqrt{npq}}\right) +$$

$$+ (nq - \lambda \sqrt{npq} + \tfrac{1}{2}) \log \left(1 - \frac{\lambda p}{\sqrt{npq}}\right) - \frac{1}{12}\left(\frac{\theta_1}{n} - \frac{\theta_2}{n} - \frac{\theta_3}{n-\nu}\right).$$

We now suppose that λ is situated between the two fixed limits λ_1 and λ_2. For all sufficiently large values of n, both $|\lambda q|/\sqrt{npq}$ and $|\lambda p|/\sqrt{npq}$

will then certainly be smaller than $\frac{1}{4}$. However, according to TAYLOR's theorem, we have for $|x| < \frac{1}{4}$

$$\log (1+x) = x - \tfrac{1}{2} x^2 + \theta x^3,$$

where $|\theta| < 1$. Applying this formula to the above expression of the quantity z, we obtain after some reductions

$$z = \frac{\lambda^2}{2} + \frac{r}{\sqrt{n}},$$

where $|r|$ is smaller than a quantity independent of n and λ. By a simple calculation, it now follows that

$$(6.4.5) \qquad p_\nu = \frac{1}{\sqrt{2 \pi n p q}} e^{-z} = \frac{1}{\sqrt{2 \pi n p q}} e^{-\frac{\lambda^2}{2}} + \frac{R}{n}$$

where $|R|$ is again smaller than a quantity independent of n and λ.

The relation 6.4.5 gives an asymptotic expression of the probability p_ν for large values of n. According to 6.4.2, it follows from this expression that we have

$$P(\lambda_1 < \lambda \leq \lambda_2) = \frac{1}{\sqrt{2 \pi n p q}} \sum e^{-\frac{\lambda^2}{2}} + \frac{1}{n} \sum R,$$

where both sums should be extended over all values of ν satisfying 6.4.3. The number of these values of ν is at most equal to $(\lambda_2 - \lambda_1) \sqrt{n p q} + 1$, and hence the last term in the second member tends to 0 as $n \to \infty$. On the other hand, the first term in the second member contains the sum of a sequence of values of the function $e^{-\lambda^2/2}$, for values of the variable situated in the interval $\lambda_1 < \lambda \leq \lambda_2$. The distance between two successive values of λ is, according to 6.4.4, equal to $1/\sqrt{n p q}$. It then follows from the definition of a definite integral that, for infinitely increasing values of n, the corresponding term tends to the limit

$$\frac{1}{\sqrt{2 \pi}} \int_{\lambda_1}^{\lambda_2} e^{-\frac{u^2}{2}} d u.$$

We have thus obtained the following highly important relation, which was first found by DE MOIVRE, and later independently deduced by LAPLACE, and which contains BERNOULLI's theorem as a particular case:

$$(6.4.6) \qquad \lim_{n \to \infty} P(\lambda_1 < \lambda \leq \lambda_2) = \frac{1}{\sqrt{2 \pi}} \int_{\lambda_1}^{\lambda_2} e^{-\frac{u^2}{2}} d u.$$

In terms of our original variable, the frequency f, this relation becomes

$$(6.4.7) \quad \lim_{n\to\infty} P\left(np+\lambda_1\sqrt{npq}<f\leqq np+\lambda_2\sqrt{npq}\right)=\frac{1}{\sqrt{2\pi}}\int_{\lambda_1}^{\lambda_2}e^{-\frac{u_2}{2}}du.$$

Ex. — If we make $n=10{,}000$ throws with a coin, the probability that the number f of heads will fall between the limits $4{,}850<f\leqq5{,}150$ will, according to 6.4.7, be approximately equal to

$$\frac{1}{\sqrt{2\pi}}\int_{-3}^{3}e^{-\frac{u^2}{2}}du=0.9973.$$

In fact, taking $p=q=\frac{1}{2}$, we obtain from 6.4.4 the limits $\lambda_1=-3$ and $\lambda_2=3$. The numerical value of the integral is then obtained from Table I.

The opposite event, i.e., the event that the number of heads will fall outside the limits given above, has a probability approximately equal to $1-0.9973=0.0027$.

For any fixed value of n, the random variable λ has a discrete distribution, with the possible values $(v-np)/\sqrt{npq}$, where $v=0,1,\ldots,n$, the corresponding probabilities p_v being given by the binomial formula. In fact, the distribution of λ is obtained from the original binomial distribution of the frequency f, if the scale of the variable is transformed by placing the origin in the mean value np, and measuring the variable in units of the standard deviation $\sqrt{npq}$. The probability that the variable λ will assume a value belonging to an arbitrarily given interval (λ_1, λ_2) is then identical with the sum of the probabilities p_v for all possible values of v such that the corresponding λ belongs to the interval. By 6.4.6 this probability tends, as $n\to\infty$, to the integral of a certain continuous function, extended over the interval (λ_1, λ_2). *We thus see that, as $n\to\infty$, while the probability p is kept constant, the discrete distribution of λ (the "standardized binomial distribution") is transformed into a continuous distribution.*

We now introduce the notations

$$\Phi(x)=\frac{1}{\sqrt{2\pi}}\int_{-\infty}^{x}e^{-\frac{t^2}{2}}dt, \qquad \varphi(x)=\Phi'(x)=\frac{1}{\sqrt{2\pi}}e^{-\frac{x^2}{2}}.$$

Then 6.4.6 can be written

$$\lim_{n\to\infty} P(\lambda_1<\lambda\leqq\lambda_2)=\int_{\lambda_1}^{\lambda_2}\varphi(x)\,dx=\Phi(\lambda_2)-\Phi(\lambda_1),$$

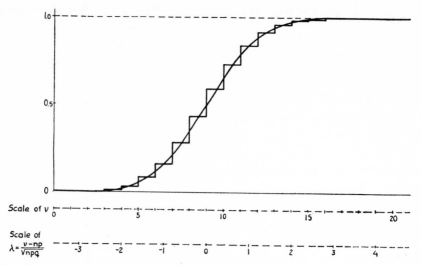

Fig. 13. Distribution functions for a binomial distribution and the approximating normal distribution. $p = 0.3$, $n = 30$.

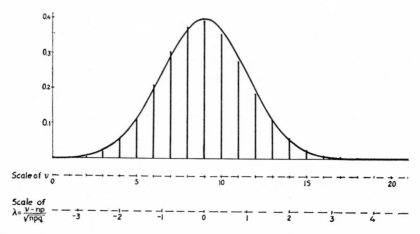

Fig. 14. Probability diagram for the same binomial distribution as in fig. 13, with the ordinates multiplied with $\sqrt{npq}$, together with the corresponding normal frequency curve.

and it is thus seen that $\varphi(x)$ is the frequency function, and consequently $\Phi(x)$ the distribution function, of the limiting distribution of the variable λ. From Appendix, 3, it follows that

$$\int_{-\infty}^{\infty} \varphi(x)\, dx = 1,$$

so that $\varphi(x)$ has all the characteristic properties (cf. 5.3) of a frequency function.

The distribution defined by the frequency function $\varphi(x)$ and the distribution function $\Phi(x)$ is known as the *normal distribution*. This distribution, which is of a fundamental importance in the theory of probability and many of its applications, will be more thoroughly discussed in the following chapter. The result obtained above can now be briefly resumed as follows:

As n tends to infinity, while the probability p is kept constant, the stand-ardized binomial distribution tends to the normal distribution as a limit.

This passage to the limit can be graphically illustrated by drawing in the same diagram the distribution curve of a λ distribution for a moderately large value of n, and the limiting curve $y = \Phi(x)$. From 6.4.5 we further find that, if the ordinates in the probability diagram of a λ distribution are multiplied by the factor $\sqrt{npq}$, we shall obtain a diagram where the upper end-points of the ordinates will, for large values of n, approximate the curve $y = \varphi(x)$. Examples of diagrams of these types are given in Figures 13–14.

It follows from the above proof that, in order that the two distributions should lie close to one another, it must be required that npq should be large. In practice, a fairly satisfactory approximation will usually be obtained as soon as $npq > 10$.

We shall finally justify the statement made above that BERNOULLI's theorem is contained as a particular case in the limiting relation 6.4.6. To this end, it is only necessary to observe that the inequality $\left|\dfrac{f}{n} - p\right| \leq \varepsilon$ is equivalent to

$$|\lambda| \leq \varepsilon \sqrt{\frac{n}{pq}}.$$

If A is any given positive number, we can always choose n so large that we have $\varepsilon \sqrt{\dfrac{n}{pq}} > A$. By the addition rule 3.2.1 a we then have

$$P\left(\left|\frac{f}{n} - p\right| \leq \varepsilon\right) \geq P(|\lambda| \leq A).$$

As $n \to \infty$, the second member tends to the limit $\displaystyle\int_{-A}^{A} \varphi(t)\,dt$, and by choosing A sufficiently large this limit can be made to lie as close to 1 as we

101

please. Hence $P\left(\left|\dfrac{f}{n}-p\right|\leqq\varepsilon\right)$ tends to 1, and the opposite probability, $P\left(\left|\dfrac{f}{n}-p\right|>\varepsilon\right)$, tends to 0, which is exactly the assertion contained in BERNOULLI's theorem.

6.5. The Poisson distribution. — In the preceding section, we have studied the behaviour of the standardized binomial distribution when the number n is allowed to increase indefinitely, while the probability p is kept constant, and we have seen that the discrete binomial distribution will then be transformed into a continuous limiting distribution, the normal distribution. By arranging the passage to the limit a little differently, we can also obtain another interesting limiting distribution which is, however, of the discrete type.

Consider, for a fixed value of ν, the general expression 6.2.1 of the probability p_ν in the binomial distribution. After some easy transformations, we may write this in the form

$$p_\nu=\binom{n}{\nu}p^\nu q^{n-\nu}=\frac{(np)^\nu}{\nu!}\left(1-\frac{np}{n}\right)^n\frac{\left(1-\dfrac{1}{n}\right)\left(1-\dfrac{2}{n}\right)\cdots\left(1-\dfrac{\nu-1}{n}\right)}{(1-p)^\nu}.$$

Let us suppose that the probability p has a very small value. Let, further, λ denote a given positive constant (note that the letter λ has here a quite different significance from the one used in the preceding section). We can now always choose the positive integer n such that np does not differ from λ by more than at most $\tfrac{1}{2}p$, and since p is very small, it will be necessary to choose n very large in order to have this condition satisfied. The above expression shows that, for these values of p and n, the probability p_ν will be approximately equal to

$$\frac{\lambda^\nu}{\nu!}\left(1-\frac{\lambda}{n}\right)^n.$$

Further, since n is very large, it is well known that $\left(1-\dfrac{\lambda}{n}\right)^n$ is approximately equal to $e^{-\lambda}$. It follows that, for the values of p and n here considered, the probability p_ν will be approximately equal to a quantity π_ν given by the expression

(6.5.1)
$$\pi_\nu=\frac{\lambda^\nu}{\nu!}e^{-\lambda}.$$

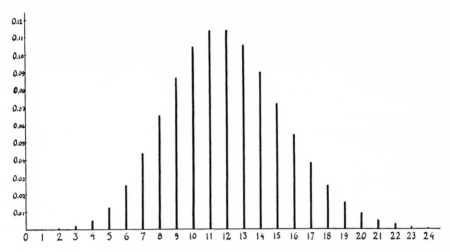

Fig. 15. The POISSON distribution. Probability diagram for $\lambda = 12$.

In the limit, when we allow p to tend to zero, while at the same time n tends to infinity in such a way that np tends to the given constant λ, the above argument shows that we have for every fixed ν the exact limiting relation

$$\lim_{\substack{p \to 0 \\ n \to \infty \\ np \to \lambda}} p_r = \pi_\nu = \frac{\lambda^\nu}{\nu!} e^{-\lambda}.$$

At the limit passage considered here, the binomial distribution thus tends to a limiting distribution of the discrete type, where the possible values form the infinite sequence $\nu = 0, 1, 2, \ldots$, the probability corresponding to any value ν being equal to π_ν. The sum of all the probabilities π_ν is easily seen to have the correct value 1, since we have

$$\sum_{0}^{\infty} \pi_\nu = e^{-\lambda} \sum_{0}^{\infty} \frac{\lambda^\nu}{\nu!} = e^{-\lambda} \cdot e^{\lambda} = 1.$$

The discrete distribution defined by the probabilities π_ν is known as the POISSON distribution. The constant parameter λ may assume any positive value. A probability diagram of the distribution for the particular case $\lambda = 12$ is shown in fig. 15.

The generating function of the POISSON distribution is according to 5.10.2

$$\psi(t) = \sum_{0}^{\infty} \pi_\nu e^{\nu t} = e^{-\lambda} \sum_{0}^{\infty} \frac{(\lambda e^t)^\nu}{\nu!} = e^{\lambda(e^t - 1)}.$$

From the derivatives of $\psi(t)$ we can find the moments of the distribution as shown by 5.10.4, and in particular we find in this way $\alpha_1 = \lambda$, $\alpha_2 = \lambda + \lambda^2$. It follows that the mean m and the variance σ^2 of the distribution are

$$m = \alpha_1 = \lambda,$$
$$\sigma^2 = \alpha_2 - \alpha_1^2 = \lambda.$$

If two independent variables have POISSON distributions with the parameter values λ_1 and λ_2, the sum of the variables has according to 5.10.7 the generating function

$$e^{\lambda_1(e^t-1)} \cdot e^{\lambda_2(e^t-1)} = e^{(\lambda_1+\lambda_2)(e^t-1)},$$

This is, however, the generating function of a POISSON distribution with the parameter $\lambda_1 + \lambda_2$. The argument is immediately extended to more than two variables. *We thus find that the POISSON distribution reproduces itself at the addition of independent variables, the parameter values corresponding to the terms being added to form the parameter of the sum.*

Ex. 1. — In certain classes of applications we are concerned with a very large number of independent experiments, in each of which there is a very small probability for the occurrence of a given event A. From the above deduction of the POISSON distribution, it may then be surmised that the total number of occurrences of A will be a variable, the distribution of which is at least approximately of the POISSON form.

Suppose, e.g., that we take a small sample from a liquid containing a large number of microorganisms in homogeneous suspension. We can then say that, for every individual microorganism contained in the liquid, we perform a random experiment with the possible outcomes "included" and "not included" in our sample, the probability of the outcome "included" being very small. We are thus led to form the hypothesis that the total number of microorganisms included in a small sample of given volume will have a POISSON distribution. In many cases it has been found that this hypothesis agrees well with observed statistical data. An example will be given at a later stage (cf. 15.2, ex. 1), after we have discussed some general methods for testing a statistical hypothesis of the kind encountered here. — The fact that data of this type often adapt themselves well to the POISSON distribution renders this distribution an important tool for many biological and medical investigations.

Ex. 2. — In a further important class of applications we are concerned with events that are randomly distributed in time. We then observe the occurrence of certain events, which may be, e.g., disintegrations of atoms from some radioactive substance, calls on a telephone line, claims in an insurance company, automobile accidents on a road etc. Let our observations begin at the time $t = 0$, and let us consider the number of events occurring during any time interval as a random variable. It is required to find the probability $p_\nu(t)$ that, up to the time t, exactly ν events will occur.

In order to solve this problem, we shall introduce two simplifying assumptions. In the first place we shall assume that the number of events occurring during a given time interval is always independent of the number of those events that have already occurred at the beginning of the interval. Moreover we shall assume that, for an infinitesimal interval $(t, t + dt)$, we have the probability $\lambda \, dt$ that exactly one event will occur, where λ is a constant, while the probability of more than one event is of a higher order of smallness than dt.

We shall first find the probability $p_0 (t)$ that there will be no event at all up to the time t. Consider the interval $(0, t + dt)$, which is composed of the two parts $(0, t)$ and $(t, t + dt)$. The probability that there will be no event in $(0, t)$ is $p_0 (t)$, and the corresponding probability for $(t, t + dt)$ is $1 - \lambda dt$ (up to terms of the order dt), so that a straightforward application of the multiplication theorem for independent events (3.5.1) gives

$$p_0 (t + dt) = p_0 (t) (1 - \lambda \, dt).$$

Since dt is regarded as infinitely small, this yields

$$p_0' (t) = - \lambda \, p_0 (t),$$

$$p_0 (t) = e^{-\lambda t},$$

where the constant of integration has been determined from the obvious condition $p_0 (0) = 1$.

For $\nu > 0$ we obtain by a similar argument, based on the addition and multiplication theorems,

$$p_\nu (t + dt) = p_\nu (t) (1 - \lambda \, dt) + p_{\nu-1} (t) \cdot \lambda \, dt,$$

$$p_\nu' (t) = - \lambda \, p_\nu (t) + \lambda \, p_{\nu-1} (t),$$

and further, writing $p_\nu (t) = r_\nu (t) e^{-\lambda t}$,

$$r_\nu' (t) = \lambda \, r_{\nu-1} (t).$$

Now $r_0 (t) = 1$, and by successive integrations we can find $r_1 (t)$, $r_2 (t)$, ..., the constants of integration being determined by the condition $r_\nu (0) = p_\nu (0) = 0$ for $\nu > 0$. In this way we find the general expression

$$(6.5.2) \qquad p_\nu (t) = \frac{(\lambda t)^\nu}{\nu !} e^{-\lambda t}.$$

Thus the number of events occurring up to the time t has a POISSON distribution with the parameter λt. With a remarkably high degree of approximation, this result is applicable, e.g., to the number of disintegrated radioactive atoms, the number of claims in certain branches of insurance, the number of accidents under various conditions etc.

According to 6.5.2 the probability that there will be at least one event before the time t is $1 - e^{-\lambda t}$. In other words, this is the probability that the first event will occur before the time t, or the d.f. of the time elapsed from the beginning of the observations until the occurrence of the first event. The corresponding fr. f. is $\lambda e^{-\lambda t}$, and the mean value is $\int_0^\infty \lambda t \, e^{-\lambda t} \, dt = 1/\lambda$. This expression gives, e.g., the "mean duration of life" of a radioactive atom, when it is assumed that the probability of a disintegration within the time element dt is $\lambda\, dt$.

The present example represents a simple particular case of a process which develops in time and is subject to random influences. In general, such a process is denoted as a *random* or *stochastic process*. For the theory of more general classes of such processes, we refer to Doob's *Stochastic Processes*, or to Feller's *Probability Theory*, Chs. 14–17.

Ex. 3. — The application of 6.5.2 to the number of claims in an insurance business forms the starting point of the *collective risk theory*, which has important applications to various actuarial problems. In this case it is convenient to introduce a slight generalization of the above considerations, choosing the *accumulated risk premium* as independent variable instead of the time. Suppose that during the time element dt the insurance company receives from its policyholders a total amount of risk premium dP, which is equal to the mean value of the total amount of claims occurring during the time element. Let us consider only the simple particular case when the constant sum 1 has to be paid for every claim. Then a trivial modification of the deduction of 6.5.2 will show that we have the probability

$$\frac{P^{\nu}}{\nu!}\, e^{-P}$$

that exactly ν claims will arise during a time interval when the total risk premium amount P is received. In practice, the amounts paid will vary from one claim to another, and it is shown in works on risk theory how similar, but more complicated expressions may be obtained for the probabilities corresponding to the general case. For a further development of this subject, we refer to a number of papers in the *Skandinavisk Aktuarietidskrift* by F. Lundberg, Cramér, Segerdahl and other authors.

Ex. 4. — In order to give an example of the application of the Poisson formula in the theory of telephone traffic, we shall now briefly indicate the deduction of an important formula in this field due to Erlang. Let us assume that n lines are available for the traffic between two stations, and that the occurrence of calls satisfies the conditions stated in Ex. 2 above. The constant λ then represents the intensity of incoming traffic, and $1/\lambda$ is the mean length of the time interval between two consecutive calls. We further assume that, if a certain line is busy at time t, the probability that it will become free during the following time element dt is $\mu\, dt$, where μ is a constant. By the same argument as in Ex. 2, it then follows that the fr. f. of the duration of a telephone conversation is $\mu\, e^{-\mu t}$, while the mean duration of a conversation is $1/\mu$. Finally, we assume that there is no "waiting line". Thus a call occurring at a moment when all n lines are busy will not be satisfied, and will have to be repeated later on.

Suppose now that all n lines are free at time $t=0$. We want to find the probability $p_{\nu}(t)$ that at time t exactly ν lines are busy, for $\nu=0,1,\ldots,n$. By arguments similar to those used in Ex. 2 we obtain for $0<\nu<n$

$$p_{\nu}(t+dt)=p_{\nu}(t)\,(1-\lambda\, dt)\,(1-\nu\,\mu\, dt)+p_{\nu-1}(t)\cdot\lambda\, dt+p_{\nu+1}(t)\,(\nu+1)\,\mu\, dt,$$

whence

$$p_{\nu}'(t)=-(\lambda+\nu\,\mu)\,p_{\nu}(t)+\lambda\,p_{\nu-1}(t)+(\nu+1)\,\mu\,p_{\nu+1}(t).$$

For $\nu = 0$ we have the same equation without the term in $p_{\nu-1}$. For $\nu = n$ it is the term in $p_{\nu+1}$ that drops out, and at the same time the coefficient of p_ν is reduced to $-n\mu$.

Thus the $n+1$ unknown functions $p_0(t), \ldots, p_n(t)$ satisfy a system of linear differential equations with constant coefficients. We shall here assume that the general method for the solution of such a system is known by the reader. For the system of equations with which we are concerned here it can be shown without difficulty that the so called characteristic equation has one root equal to zero, while the n remaining roots are simple and negative. It follows that every solution of the system of differential equations can be written in the form

$$p_\nu(t) = \pi_\nu + c_{\nu 1} e^{-r_1 t} + \cdots + c_{\nu n} e^{-r_n t},$$

where $r_1, \ldots, r_n$ are all positive. When t tends to infinity, we thus obtain

$$\lim_{t \to \infty} p_\nu(t) = \pi_\nu, \qquad \lim_{t \to \infty} p_\nu'(t) = 0.$$

Thus for large values of t the telephone traffic considered here will approach a state of statistical equilibrium, characterized by the fact that all probabilities $p_\nu(t)$ tend to constant limiting values. If we allow t to tend to infinity in the differential equations given above, we obtain a system of ordinary linear and homogeneous equations for the unknown quantities $\pi_0, \ldots, \pi_n$. Writing $\lambda/\mu = \varkappa$, these equations become

$$\pi_1 = = \varkappa \pi_0,$$

$$2\pi_2 = (\varkappa + 1)\pi_1 - \varkappa \pi_0 = \varkappa \pi_1,$$

$$\cdot \ \cdot \ \cdot \ \cdot \ \cdot \ \cdot \ \cdot \ \cdot \ \cdot \ \cdot \ \cdot \ \cdot \ \cdot \ \cdot$$

$$n\pi_n = (\varkappa + n - 1)\pi_{n-1} - \varkappa \pi_{n-2} = \varkappa \pi_{n-1}.$$

Hence we obtain

$$\pi_\nu = \frac{\varkappa^\nu}{\nu!} \pi_0.$$

Finally, the condition $\sum_0^n p_\nu(t) = 1$ gives $\sum_0^n \pi_\nu = 1$, and we thus obtain ERLANG's formula

$$\pi_\nu = \lim_{t \to \infty} p_\nu(t) = \frac{\dfrac{\varkappa^\nu}{\nu!}}{1 + \dfrac{\varkappa}{1!} + \dfrac{\varkappa^2}{2!} + \cdots + \dfrac{\varkappa^n}{n!}}.$$

In the particular case when $n = \infty$, we have the POISSON expression $\pi_\nu = \dfrac{\varkappa^\nu}{\nu!} e^{-\varkappa}$.

6.6. Problems. — 1. Deduce the following inequalities for the "tails" of the binomial distribution:

$$\sum_{0}^{k} \binom{n}{\nu} p^{\nu} q^{n-\nu} < \frac{(n-k+1)\,p}{n\,p+p-k} \cdot \binom{n}{k} p^{k} q^{n-k} \text{ for } 0 < k < n\,p+p,$$

$$\sum_{k}^{n} \binom{n}{\nu} p^{\nu} q^{n-\nu} < \frac{(k+1)\,q}{k-(n\,p-q)} \cdot \binom{n}{k} p^{k} q^{n-k} \text{ for } n\,p-q < k < n.$$

2. The number of heads obtained in 24 throws with a coin is denoted by X. Compute to five decimal places the probability $P\left(\left| X - E(X) \right| > k\,D(X)\right)$ for $k = 1, 2, \ldots, 5$. Compare the result with the approximation obtained under the assumption that X is normally distributed, and also with the upper bounds supplied by TCHEBYCHEFF's theorem.

3. Use 6.4.5 to study the asymptotic behaviour of the probability π_n considered in 4.1, ex. 2.

4. A certain unusual event A has the probability $1/n$, where n is large. The probability that A will not occur in a series of k independent observations is $(1-1/n)^{k}$. When k is a small number in comparison with n, this is approximately equal to one, so that we can be practically certain that the event will not occur (cf. 3.1). On the other hand, when k is large, $(1-1/n)^{k}$ may differ very considerably from 1. Find an approximate value in the particular case when $k = n$. Would you think it safe to say that we can be practically certain that A will not occur in a series of $k = n$ observations?

5. For a certain microorganism, the probability of splitting in two organisms during the time element dt is $\lambda\,dt$, where λ is a constant. At time $t = 0$ one single organism is present. Let $p_n(t)$ denote the probability that there are exactly n organisms at time t. Show that $p_n(t) = e^{-\lambda t}(1 - e^{-\lambda t})^{n-1}$ and that $E(n) = e^{\lambda t}$. Generalize this to the case when there are any given number n_0 of organisms at time $t = 0$.

CHAPTER 7

THE NORMAL DISTRIBUTION

7.1. The normal functions. — In 6.4 we have already introduced the *normal distribution function*

$$\Phi(x) = \frac{1}{\sqrt{2\,\pi}} \int_{-\infty}^{x} e^{-\frac{t^{2}}{2}}\,dt,$$

and the *normal frequency function*

$$\varphi(x) = \Phi'(x) = \frac{1}{\sqrt{2\,\pi}} e^{-\frac{x^{2}}{2}}.$$

Diagrams of these functions are shown in Figs. 16–17, while some numerical values are given in Table I.

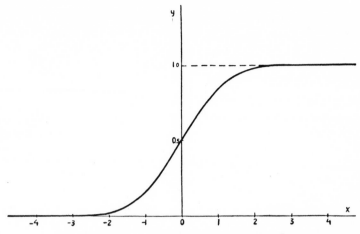

Fig. 16. The normal distribution function $\Phi(x)$.

The distribution is symmetric about the origin, and it follows (cf. 5.6) that the mean and all central moments of odd orders are zero. In order to find the central moments of even orders we start from the relation (cf. Appendix, 3)

$$\int_{-\infty}^{\infty} e^{-\frac{1}{2}h x^2} dx = \sqrt{\frac{2\pi}{h}},$$

and we differentiate this n times with respect to h. It is legitimate to differentiate under the sign of integration, and if we put $h = 1$ after the operation, we obtain the following expression for the central moment of order $2n$:

$$\mu_{2n} = \frac{1}{\sqrt{2\pi}} \int_{-\infty}^{\infty} x^{2n} e^{-\frac{x^2}{2}} dx = 2^n \cdot \frac{1}{2}\frac{3}{2} \cdots \frac{2n-1}{2} = 1 \cdot 3 \cdots (2n-1).$$

For $n = 1$ and $n = 2$ this gives

$$\mu_2 = \sigma^2 = 1, \qquad \mu_4 = 3.$$

Thus the standard deviation is equal to unity. For the measures of skewness and excess we obtain according to 5.7.8 and 5.7.9

$$\gamma_1 = \frac{\mu_3}{\sigma^3} = 0, \qquad \gamma_2 = \frac{\mu_4}{\sigma^4} - 3 = 0.$$

109

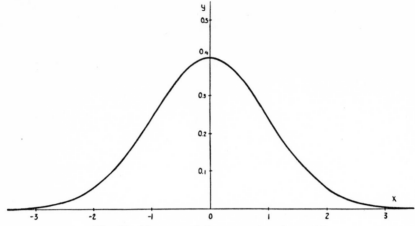

Fig. 17. The normal frequency function $\varphi(x)$.

The symmetric character of the distribution may be expressed by the following relation which is often useful;

$$\Phi(x) + \Phi(-x) = 1.$$

7.2. The normal distribution. — If a random variable X has the d.f. $\Phi\left(\dfrac{x-m}{\sigma}\right)$, where $\sigma > 0$ and m are constants, we shall say that X has a *normal distribution*. The fr.f. of X is then

$$\frac{1}{\sigma}\varphi\left(\frac{x-m}{\sigma}\right) = \frac{1}{\sigma\sqrt{2\pi}}e^{-\frac{(x-m)^2}{2\sigma^2}}.$$

The mean and the variance of X are

$$E(X) = \frac{1}{\sigma}\int_{-\infty}^{\infty} x\,\varphi\left(\frac{x-m}{\sigma}\right)dx = \int_{-\infty}^{\infty}(m+\sigma x)\,\varphi(x)\,dx = m,$$

$$D^2(X) = \frac{1}{\sigma}\int_{-\infty}^{\infty}(x-m)^2\,\varphi\left(\frac{x-m}{\sigma}\right)dx = \sigma^2\int_{-\infty}^{\infty}x^2\,\varphi(x)\,dx = \sigma^2,$$

which shows that the parameters m and σ have their usual significance as the mean and the standard deviation of the distribution. Thus if we say that X is normally distributed with mean m and s.d. σ, the distribution will be completely characterized. For the sake of brevity we shall in this

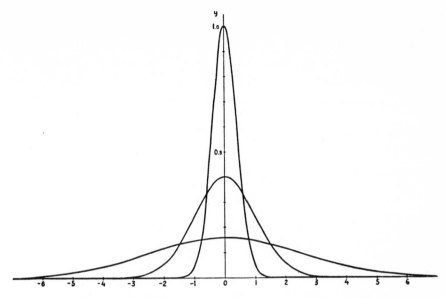

Fig. 18. Normal frequency curves. $m=0$, $\sigma=0.4$, 1.0, 2.5.

case simply say that X *is normal* (m, σ). The corresponding standardized variable $(X-m)/\sigma$ will then have the *standardized normal distribution*, with the d.f. $\Phi(x)$ and the fr.f. $\varphi(x)$.

In the general case the normal distribution is symmetric about the maximum point of the fr.f., $x=m$, where the mean, the median and the mode of the distribution coincide. All odd central moments are zero, while in the same way as for the variance we find

$$(7.2.1) \qquad\qquad \mu_{2n} = 1 \cdot 3 \cdot \ldots (2n-1)\,\sigma^{2n}.$$

By forming the second derivative of the fr.f. it will be seen that the frequency curve

$$y = \frac{1}{\sigma\sqrt{2\,\pi}}\,e^{-\frac{(x-m)^2}{2\sigma^2}}$$

has two symmetric points of inflexion at $x=m\pm\sigma$. The total area of the region included between the curve and the axis of x is of course equal to 1. A change in the value of m causes only a displacement of the curve, without modifying its form, whereas a change in the value of σ amounts to a change of scale on both coordinate axes. Some normal frequency curves corresponding to $m=0$ and different values of σ are shown in Fig. 18.

The smaller we take σ, the more we concentrate the mass of the distribution in the neighbourhood of the point $x = m$. In the limiting case $\sigma = 0$, the whole mass will be concentrated at $x = m$, so that the distribution will then be a one point distribution (cf. 6.1) with the d.f. $\varepsilon (x - m)$. Thus the latter distribution can be regarded as a limiting case of the normal distribution.

If X is normal (m, σ), the probability that X will assume a value between the limits $m + \lambda_1 \sigma$ and $m + \lambda_2 \sigma$ is

$$P (m + \lambda_1 \sigma < X < m + \lambda_2 \sigma) = \frac{1}{\sigma} \int\limits_{m+\lambda_1\sigma}^{m+\lambda_2\sigma} \varphi \left(\frac{x - m}{\sigma} \right) dx$$

$$= \int\limits_{\lambda_1}^{\lambda_2} \varphi (x) \, dx = \Phi (\lambda_2) - \Phi (\lambda_1).$$

Thus any probability of this type can be easily found by means of a table of values of the function $\Phi (x)$, such as the one given in Table I.

The probability that X will assume a value satisfying the inequality $|X - m| > \lambda \sigma$ is on account of the symmetry of the distribution

(7.2.2)
$$P (|X - m| > \lambda \sigma) = P (X < m - \lambda \sigma) + P (X > m + \lambda \sigma)$$
$$= \Phi (-\lambda) + 1 - \Phi (\lambda) = 2 (1 - \Phi (\lambda)).$$

In the two last columns of Table II, this probability is tabulated as a function of λ. From this table, we find, e.g., that the probability that a normally distributed variable will deviate from its mean in either direction by more than two times the standard deviation is 4.55 %, while the probability of a deviation exceeding three times the s.d. is only 0.27 %. Thus a deviation exceeding three times the s.d. will, on the average, only occur about 27 times in 10,000 trials. This must be regarded as an *unusual event* in the sense of 3.1. Thus if we make one single or a comparatively small number of observations on a variable which is supposed to be normal (m, σ), and if in one of our observations we find a deviation from m exceeding 3σ, we shall have reason to ask ourselves if the hypothesis concerning the distribution is really well founded.

If, conversely, we want to find λ so that the probability $P (|X - m| > \lambda \sigma)$ assumes a given value, say p %, we shall have to solve the equation in λ

$$P (|X - m| > \lambda \sigma) = 2 (1 - \Phi (\lambda)) = \frac{p}{100}.$$

112

The unique root of this equation will be denoted by $\lambda = \lambda_p$, and called the p *percent value of the normal distribution*. We thus have the probability p % that any normally distributed variable will deviate from its mean by more than λ_p times the s.d. Such p percent values are extensively used in statistical practice (cf. Ch. 14). In the two first columns of Table II, λ_p is tabulated as a function of p. We find here, e.g., that the 10 % value is 1.64, which means that we have the probability 10 % of a deviation exceeding 1.64 times the s.d.

Finally we observe that, if the variable X is normal (m, σ), it follows from 5.3.5 that any linear function $aX + b$ will also be normally distributed, with mean $am + b$ and s.d. $|a|\sigma$.

7.3. Addition of independent normal variables. — Let $X_1, \ldots, X_n$ be independent variables, such that X_ν is normal (m_ν, σ_ν). *We shall now prove the important theorem that the sum*

$$X = X_1 + \cdots + X_n$$

is also normally distributed, with mean m and standard deviation σ given by the expressions

$$m = m_1 + \cdots + m_n,$$

$$\sigma^2 = \sigma_1^2 + \cdots + \sigma_n^2.$$

Thus the normal distribution, like the POISSON distribution (cf. 6.5), has the property of reproducing itself at the addition of independent variables.

The expressions given for m and σ follow directly from 5.9.1, and hold for independent variables irrespective of their distributions. Consequently we have only to prove that the sum X has a normal distribution, and it is clearly sufficient to prove this for the particular case $n = 2$, the extension to an arbitrary n being then immediate.

If the independent variables X_1 and X_2 are normal (m_1, σ_1) and (m_2, σ_2) respectively, the sum $X = X_1 + X_2$ will according to 5.9.3 have the frequency function

$$\frac{1}{\sigma_1 \sigma_2} \int_{-\infty}^{\infty} \varphi\left(\frac{x - t - m_1}{\sigma_1}\right) \varphi\left(\frac{t - m_2}{\sigma_2}\right) dt = \frac{1}{2\pi\sigma_1\sigma_2} \int_{-\infty}^{\infty} e^{-\frac{1}{2}Q} dt,$$

where after some calculation it is seen that Q can be written in the form

$$Q = \frac{(x - t - m_1)^2}{\sigma_1^2} + \frac{(t - m_2)^2}{\sigma_2^2} = \frac{\sigma_1^2 + \sigma_2^2}{\sigma_1^2 \sigma_2^2}\left(t - \frac{\sigma_1^2 m_2 + \sigma_2^2 (x - m_1)}{\sigma_1^2 + \sigma_2^2}\right)^2 + \frac{(x - m_1 - m_2)^2}{\sigma_1^2 + \sigma_2^2}.$$

113

Writing here m for $m_1 + m_2$ and σ^2 for $\sigma_1^2 + \sigma_2^2$, and introducing the substitution

$$u = \frac{\sigma}{\sigma_1 \sigma_2} \left(t - \frac{\sigma_1^2 m_2 + \sigma_2^2 (x - m_1)}{\sigma^2} \right),$$

the above expression for the frequency function will take the form

$$\frac{1}{2 \pi \sigma} e^{-\frac{(x-m)^2}{2\sigma^2}} \int_{-\infty}^{\infty} e^{-\frac{u^2}{2}} d u = \frac{1}{\sigma \sqrt{2\pi}} e^{-\frac{(x-m)^2}{2\sigma^2}},$$

so that our theorem is proved.

The same theorem can also be proved by means of generating functions. By 5.10.3, the generating function of a variable which is normal (m, σ) is

$$\psi (t) = \int_{-\infty}^{\infty} e^{t x} \cdot \frac{1}{\sigma \sqrt{2\pi}} e^{-\frac{(x-m)^2}{2\sigma^2}} d x.$$

After some reductions, this expression takes the form

$$\psi (t) = e^{mt + \frac{1}{2} \sigma^2 t^2},$$

and it is now evident that this generating function will reproduce itself by multiplication, the parameters m and σ^2 being added in the way stated in our theorem.

Using the last remark of 7.2, we obtain the more general result that any linear function $a_1 X_1 + \cdots + a_n X_n + b$ of independent normal variables is itself normally distributed.

An interesting particular case of a linear function is the arithmetic mean $\bar{X} = (X_1 + \cdots + X_n)/n$. Suppose that all the X_i are independent and normal (m, σ). Then (cf. 5.7.6 and 5.7.7) $\bar{X}$ will be normal $(m, \sigma/\sqrt{n})$. For large n the s.d. becomes small, so that the bulk of the mass in the distribution of $\bar{X}$ will be concentrated near the point m.

The generalization of BERNOULLI's theorem given in 6.3 shows that the last remark applies also to any non-normal distribution, as long as the s.d. σ is finite. On the other hand, we have seen in 5.9, ex. 2, that for the CAUCHY distribution, where m and σ do not exist, the arithmetic mean has the same distribution as any of the X_ν, so that there is no concentration of the distribution about any particular point for large n.

7.4. The central limit theorem. — If $X_1, \ldots, X_n$ are independent and normally distributed variables, we have seen in the preceding section that the sum

$$X = X_1 + \cdots + X_n$$

is itself normally distributed. The standardized variable $(X-m)/\sigma$ then has the d.f. $\Phi(x)$ and the fr.f. $\varphi(x)$.

Let us now suppose that the variables $X_1, \ldots, X_n$ are independent, but *not* necessarily normal. Will it then be possible to make any general statement about the distribution of the sum X?

In 6.4 we have already solved this question in a particular case. Consider the case when each X_ν has the only possible values 1 and 0, the corresponding probabilities being p and $q = 1 - p$ respectively. Then according to 6.2 the sum X will have a binomial distribution, with mean np and s.d. $\sqrt{npq}$. The corresponding standardized variable is in this case $(X-np)/\sqrt{npq}$, and by DE MOIVRE's theorem (cf. 6.4) the distribution of this variable tends for infinitely increasing n to the standardized normal distribution, with the d.f. $\Phi(x)$.

Now a very important theorem, known as the *central limit theorem* of the theory of probability, states that the same property holds true under much more general conditions about the distributions of the X_ν. Suppose, as before, that X_ν has mean m_ν and s.d. σ_ν, so that the mean and the variance of the sum X are given by the usual expressions

$$m = m_1 + \cdots + m_n, \qquad \sigma^2 = \sigma_1^2 + \cdots + \sigma_n^2.$$

Then the central limit theorem asserts that, under very general conditions with respect to the distributions of the X_ν, the d.f. of the standardized sum $(X-m)/\sigma$ will for large n be approximately equal to $\Phi(x)$. If the sequence of the variables X_ν is unlimited, we can even assert that we have

$$(7.4.1) \qquad \lim_{n \to \infty} P\left(\frac{X-m}{\sigma} \leq x\right) = \Phi(x).$$

In such a case we shall say that the variable X is *asymptotically normal* (m, σ).

This theorem has been stated by LAPLACE in his classical treatise of 1812. A rigorous proof under fairly general conditions, however, was not given until 1901, by the Russian mathematician LIAPOUNOFF. Since then, the theorem has been given a more precise form and the conditions of validity have been generalized by a large number of authors.

A rigorous proof of the general theorem requires somewhat intricate mathematics, and we shall here satisfy ourselves with the detailed proof of a particular case that has already been given in 6.4. We mention only as an example that a sufficient condition for the validity of 7.4.1 is that all X_ν have the same distribution. This common distribution of the X_ν

may be chosen quite arbitrarily, provided only that its first and second order moments exist. Let the mean and the s.d. of this common distribution be denoted by m_0 and σ_0. The sum X will then have the mean $n\,m_0$ and the s.d. $\sigma_0 \sqrt{n}$, and will thus be asymptotically normal ($n\,m_0$, $\sigma_0 \sqrt{n}$). It follows that the arithmetic mean

$$\bar{X} = \frac{X_1 + \cdots + X_n}{n}$$

is asymptotically normal (m_0, $\sigma_0/\sqrt{n}$). (Cf. the last remarks of the preceding section.)

In the more general case when the X_ν distributions are not necessarily equal, the conditions required to establish the validity of 7.4.1 may be broadly characterized as expressing that each single X_ν should, on the average, only give a relatively insignificant contribution to the total X.

The central limit theorem, as expressed by the relation 7.4.1, is concerned with the asymptotic behaviour as $n \to \infty$ of the *distribution function* of the standardized sum X. Under somewhat more restrictive conditions it can also be shown that, when the X_ν have distributions of the continuous type, the *frequency function* of $(X - m)/\sigma$ tends to the normal fr.f. $\varphi(x)$.

The central limit theorem is concerned with one particular symmetric function of a large number of variables $X_1, \ldots, X_n$, viz. the sum of the variables. It is possible to generalize the theorem to other kinds of symmetric functions, such as the sum of the k:th powers of the variables, a rational function of a certain number of sums of powers, etc. Under fairly general conditions it can be shown that, when n is large, such a function is approximately normally distributed.

For a more complete discussion of the questions mentioned here and in the two following sections we must refer to more advanced works, such as *Random Variables*, Chs. 6–7, and *Mathematical Methods*, Chs. 17 and 28.

Ex. — Draw the frequency curves of the distributions of sums of two, three and four independent variables with rectangular distributions discussed in 5.9, ex. 1. Compare each curve with a normal frequency curve having the same mean and standard deviation.

7.5. Asymptotic expansion derived from the normal distribution. — Let X be a variable of the continuous type with mean m and s.d. σ, and let $f(x)$ denote the frequency function of the standardized variable $(X - m)/\sigma$. If it is known that X is the sum of a large number of independent variables of continuous type, the central limit theorem tells us that, under

116

certain general conditions, $f(x)$ will be approximately equal to the normal fr.f. $\varphi(x)$.

Under certain conditions it is even possible to find a more accurate representation of $f(x)$. Such a representation can be given in the form of an expansion starting with the normal function $\varphi(x)$ and containing in its later terms certain derivatives of $\varphi(x)$, multiplied with constant coefficients. This expansion may be regarded as an asymptotic expansion of $f(x)$ for large values of n, in the sense that the representation of $f(x)$ for large n will be improved by including more terms in the expansion. These expansions have been studied i.a. by CHARLIER, EDGEWORTH and CRAMÉR (cf. *Random Variables*, Ch. 7).

If we consider only one additional term beyond the normal function, we shall have the expression

$$f(x) = \varphi(x) - \frac{\gamma_1}{3!} \varphi^{(3)}(x),$$

where γ_1 is the measure of skewness (cf. 5.7) of the variable X. In many cases this simple expression yields a sufficiently accurate representation of $f(x)$. When it is required to include a further term in the expansion, it is often recommended in the literature that a following term containing the fourth derivative $\varphi^{(4)}(x)$ be considered. The theoretically correct rule is, however, to let the following term contain both the fourth and the sixth derivatives, thus arriving at the expansion

$$(7.5.1) \qquad f(x) = \varphi(x) - \frac{\gamma_1}{3!}\varphi^{(3)}(x) + \frac{\gamma_2}{4!}\varphi^{(4)}(x) + \frac{10\gamma_1^2}{6!}\varphi^{(6)}(x),$$

γ_2 being the measure of excess. The reason for preferring this expansion is that it can be shown that the coefficients of the terms in $\varphi^{(4)}$ and $\varphi^{(6)}$ are, under general conditions, of the same order of magnitude for large n. In practice it is usually not advisable to go beyond this expansion.

By including the additional terms containing the derivatives, the symmetry of the normal function $\varphi(x)$ is destroyed. We find, in fact, that the function represented by 7.5.1 has the same mean $x=0$ as the normal function $\varphi(x)$, while the mode is approximately $x = -\frac{1}{2}\gamma_1$, so that the mean and the mode no longer coincide when $\gamma_1 \neq 0$. It further follows from 7.5.1 that we have

$$\frac{f(0) - \varphi(0)}{\varphi(0)} = \frac{1}{8}\gamma_2 - \frac{5}{24}\gamma_1^2.$$

117

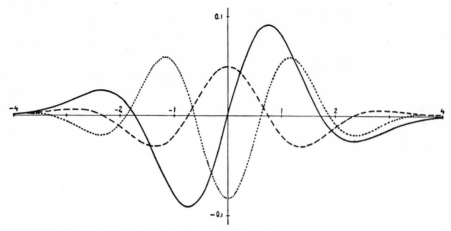

Fig. 19. Derivatives of the normal frequency function $\varphi(x)$.

$$\frac{1}{3!}\varphi^{(3)}(x) \quad \text{————————}$$

$$\frac{1}{4!}\varphi^{(4)}(x) \quad \text{— — — — — —}$$

$$\frac{10}{6!}\varphi^{(6)}(x) \quad \text{·······················}$$

Thus the relative "excess" of the frequency curve over the normal curve at $x=0$ is measured by the second member of the last relation, and this shows that the current usage of regarding γ_2 as a measure of excess is not entirely satisfactory.

Diagrams of the functions $\varphi^{(3)}$, $\varphi^{(4)}$ and $\varphi^{(6)}$, multiplied by the numerical coefficients appearing in 7.5.1, are shown in Fig. 19, while the numerical values of the functions are given in Table I.

7.6. Transformed normal distributions. — If X is a random variable such that a certain non-linear function of X is normally distributed, we shall say that X itself has a *transformed normal distribution*.

An important case of a distribution of this type is the *log-normal distribution*. A variable X is said to have a log-normal distribution if there exists a constant a such that $\log(X-a)$ is normally distributed. If $\log(X-a)$ is normal (m, σ), it is easily found (cf. 5.3) that X has the frequency function

$$(7.6.1) \qquad f(x) = \frac{1}{\sigma(x-a)\sqrt{2\pi}} e^{-\frac{(\log(x-a)-m)^2}{2\sigma^2}}$$

for $x>a$, while $f(x)=0$ for $x<a$. The corresponding frequency curve is shown in Fig. 20.

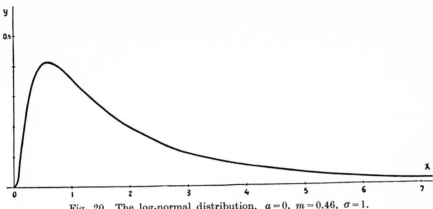

Fig. 20. The log-normal distribution. $a = 0$, $m = 0.46$, $\sigma = 1$.

The log-normal distribution appears sometimes in applications, e.g. in certain kinds of biologic and economic statistics. Suppose, e.g., that the size X of a specified organ in an observed individual may be regarded as the joint effect of a large number of mutually independent impulses $z_1 \ldots, z_n$, acting in the order of their subscripts. If these impulses simply add their effects, which are assumed to be random variables, we infer by the central limit theorem that the sum X will be asymptotically normally distributed. In general it does not, however, seem plausible that the impulses co-operate by simple addition. It seems more natural to suppose that each impulse produces an increase of the organ, the magnitude of which depends both on the strength of the impulse and on the size of the organ already attained at the moment when the impulse is working. Consider the simple case when the increase produced by the impulse z_ν is assumed to be directly proportional to z_ν and to the size X_ν resulting from the action of the $\nu - 1$ preceding impulses. We then have $X_{\nu+1} - X_\nu = k z_\nu X_\nu$, and hence

$$z_1 + \cdots + z_n = \frac{1}{k} \sum_1^n \frac{X_{\nu+1} - X_\nu}{X_\nu}.$$

If the increase produced by each single impulse is small, the last expression is approximately equal to

$$\frac{1}{k} \int_{X_1}^{X} \frac{dt}{t} = \frac{1}{k} \log \frac{X}{X_1}.$$

For large n, the sum $z_1 + \cdots + z_n$ is according to the central limit theorem approximately normally distributed, and it thus follows that X has an approximately log-normal distribution.

Similar arguments may be applied also in certain branches of economic statistics. Consider the distribution of incomes or property values in some given population. The position of an individual on the income or property scale might be regarded as the joint effect of a large number of impulses, each of which causes a certain increase of his wealth. It might be argued that the effect of a single impulse might not unreasonably be expected to be proportional to the wealth already attained. If, in addition, the impulses may be regarded as mutually independent, we should expect distributions of incomes or property values to be approximately log-normal. An example will be given in 12.3, Fig. 27. Another type of distribution which is often useful in these connections is the PARETO distribution that will be discussed in 8.4.

7.7. The importance of the normal distribution for the applications.

— In the applications of probability theory we are often concerned with random variables that may be regarded as sums of a large number of independent components, each of which contributes only an insignificant amount to the total. Thus according to the "hypothesis of elementary errors" the total error committed at a physical or astronomical measurement is regarded as produced by a large number of elementary errors, which add their effects. If often seems reasonable to apply a similar hypothesis to the variables which express the results of various kinds of industrial, biological and medical experiments. There are also other instances. Thus the total amount of claims in an insurance company is the sum of the amounts claimed on the individual policies; the total consumption of electric energy delivered by a given producer is the sum of the amounts consumed by the various customers, etc.

In all cases of this type, we should expect the random variables concerned to be at least approximately normally distributed. In fact, this conjecture is largely corroborated by statistical experience. Some examples will be discussed in Part III (cf. in particular 12.3 and 15.2).

Moreover, even in cases where there is no reason to suppose that we are dealing with normally distributed variables, current statistical methods are often based on the consideration of arithmetic means and other simple symmetric functions of large numbers of independent variables. According to the remarks made in 7.4 in connection with the central limit theorem, these functions will often be approximately normally distributed. This fact has great importance for the theory and applications of statistical methods (cf. 13.3).

In all the cases here mentioned the normal distribution constitutes an ideal limiting case, which in the applications will often be realized with a tolerable degree of approximation. Sometimes it may be possible to improve the accuracy by adding terms containing derivatives of the normal function, as shown in 7.5, or by introducing transformed normal distributions such as the log-normal distribution discussed in 7.6.

7.8. Problems.

— **1.** In a certain human population, head index i (the breadth of the cranium expressed as a percentage of the length) is assumed to be normally distributed among the individuals. There are 58 % dolichocephalics ($i \leqq 75$), 38 % mesocephalics ($75 < i \leqq 80$), and 4 % brachycephalics ($i > 80$). Find the mean and the s.d. of i.

2. The variable X is normal $(0, \sigma)$. Find $E|X|$.

3. Deduce explicit expressions for the first four derivatives of the normal fr. f. $\varphi(x)$. and find the zeros of these derivatives.

4. The variable X is normal $(0, \sigma)$. Two independent observations are made on X, and the largest of the two observed values is denoted by Z. Find the mean value of Z.

5. Add n real numbers, each of which is rounded off to the nearest integer. It is assumed that each rounding off error is a random variable uniformly distributed between $-\frac{1}{2}$ and $+\frac{1}{2}$, and that the n rounding off errors are mutually independent. a) Find the largest possible absolute value of the error in the sum due to the rounding off errors in the terms. b) Use the central limit theorem to find a quantity A such that the probability that the error in the sum will be included between the limits A is approximately 99 %. Numerical computation for $n = 10$, 100, and 1000.

6. Let a_1, a_2, and γ_1 be the first two moments and the measure of skewness of the log-normal distribution 7.6.1. Show that the equation $\eta^3 + 3\eta - \gamma_1 = 0$ has one and only one real root η, and express the parameters a, m, and σ in terms of a_1, a_2, and η.

7. Let ζ_1, ζ_2, and ζ_3 denote the quartiles (cf. 5.7) of the log-normal distribution. Express the parameters a, m, and σ in terms of ζ_1, ζ_2, and ζ_3.

CHAPTER 8

FURTHER CONTINUOUS DISTRIBUTIONS

8.1. The χ^2 distribution. — In the first three sections of this chapter, we shall study the distributions of certain simple functions of independent normal variables. All these distributions have important statistical applications, and will reappear in various connections in Part III. Thus the χ^2 distribution that will be discussed in the present section has one of its most important applications in connection with the χ^2 test, which is often used in statistical practice, and will be introduced below in Ch. 15.

Let $X_1, \ldots, X_n$ be independent random variables, each of which is normal $(0, 1)$. According to 5.4.11 the joint frequency function of these variables is

$$\frac{1}{(2\pi)^{\frac{n}{2}}} e^{-\frac{1}{2}(x_1^2 + \cdots + x_n^2)} .$$

We shall now try to find the distribution of the variable

$$\chi^2 = X_1^2 + \cdots + X_n^2 .$$

121

Evidently we always have $\chi^2 \geqq 0$. Thus if we denote the d.f. of χ^2 by $K_n(x)$, and the fr.f. by $k_n(x)$, both these functions will be equal to zero for all negative values of x. For $x > 0$, on the other hand, the d.f. $K_n(x)$ will be equal to the amount of mass in the joint distribution, which belongs to the domain $x_1^2 + \cdots + x_n^2 \leqq x$, so that we have

$$K_n(x) = P(\chi^2 \leqq x) = \frac{1}{(2\pi)^{\frac{n}{2}}} \int \cdots \int_{x_1^2 + \cdots + x_n^2 \leqq x} e^{-\frac{1}{2}(x_1^2 + \cdots + x_n^2)} dx_1 \ldots dx_n.$$

By the first mean value theorem of the integral calculus, this gives

$$K_n(x+h) - K_n(x) = \frac{1}{(2\pi)^{\frac{n}{2}}} \int \cdots \int_{x < x_1^2 + \cdots + x_n^2 \leqq x+h} e^{-\frac{1}{2}(x_1^2 + \cdots + x_n^2)} dx_1 \ldots dx_n$$

$$= \frac{1}{(2\pi)^{\frac{n}{2}}} e^{-\frac{1}{2}(x+\theta h)} (S_n(x+h) - S_n(x)),$$

where $0 < \theta < 1$, while

$$S_n(x) = \int \cdots \int_{x_1^2 + \cdots + x_n^2 \leqq x} dx_1 \ldots dx_n$$

is the "volume" of an n-dimensional sphere with the radius $\sqrt{x}$. In the sequel we shall use the letter C to denote an unspecified constant, which may depend on n but not on x, and may have different values in different formulae. By the substitution $x_\nu = y_\nu \sqrt{x}$ we obtain from the last relation

$$S_n(x) = x^{\frac{n}{2}} \int \cdots \int_{y_1^2 + \cdots + y_n^2 \leqq 1} dy_1 \ldots dy_n = C x^{\frac{n}{2}},$$

whence

$$\frac{K_n(x+h) - K_n(x)}{h} = C e^{-\frac{1}{2}(x+\theta h)} \frac{(x+h)^{\frac{n}{2}} - x^{\frac{n}{2}}}{h}.$$

Allowing h to tend to zero, we obtain

$$k_n(x) = K_n'(x) = C x^{\frac{n}{2} - 1} e^{-\frac{x}{2}}.$$

Finally the value of C in the last relation is determined from the condition

$$\int\limits_{0}^{\infty} k_n\,(x)\,d\,x = 1,$$

which gives (cf. Appendix, 2)

(8.1.1) $$k_n\,(x) = \frac{1}{2^{\frac{n}{2}}\,\Gamma\left(\frac{n}{2}\right)}\; x^{\frac{n}{2}-1}\; e^{-\frac{x}{2}}.$$

For $n=1$ and $n=2$ the frequency curve $y=k_n\,(x)$ is seen to be steadily decreasing for all positive x. For $n>2$, on the other hand, we have $y=0$ for $x=0$, and then y increases to a maximum for $x=n-2$. When x increases from $n-2$ to infinity, y decreases asymptotically to zero. A diagram of the frequency curve for the case $n=6$ is shown in Fig. 10, p. 80.

The moments of the χ^2 distribution are easily found by means of Appendix, 2, and we find

$$\alpha_\nu = n\,(n+2)\,\ldots\,(n+2\,\nu-2).$$

Hence we obtain (cf. also 8.5, ex. 1)

$$E\,(\chi^2) = \alpha_1 = n, \qquad D^2\,(\chi^2) = \alpha_2 - \alpha_1^2 = 2\,n.$$

The parameter n appearing in the χ^2 distribution represents the number of independent or "free" squares entering into the expression of χ^2. The number n is usually called the number of *"degrees of freedom"* of the distribution. The significance of this expression will appear more clearly in connection with the statistical applications of the distribution that will be discussed in Chs. 13–16.

The probability that the random variable χ^2 assumes a value exceeding a given quantity χ_0^2 is

$$P\,\{\chi^2 > \chi_0^2\} = \int\limits_{\chi_0^2}^{\infty} k_n\,(x)\,d\,x.$$

If, conversely, we want to find a quantity χ_p^2 such that the probability $P\,(\chi^2 > \chi_p^2)$ takes a given value, say $p\,\%$, we shall have to solve the equation

$$P\,(\chi^2 > \chi_p^2) = \frac{p}{100}.$$

The unique root χ_p^2 of this equation is called the *p percent value of χ^2* for *n degrees of freedom*. In Table III this value is given for some values of p and n.

The χ^2 distribution is a particular case of a more general distribution known as the Pearson type III distribution (cf. 8.4).

Ex. — Let $r = \sqrt{u^2 + v^2 + w^2}$ denote the absolute velocity of a gas molecule, while u, v and w are the components of the velocity in reference to three rectangular coordinate axes. Let us assume that, for a molecule selected at random from a given quantity of gas, u, v and w are independent random variables, each of which is normal $(0, \sigma)$. The variable

$$\left(\frac{r}{\sigma}\right)^2 = \left(\frac{u}{\sigma}\right)^2 + \left(\frac{v}{\sigma}\right)^2 + \left(\frac{w}{\sigma}\right)^2$$

will then have a χ^2 distribution with $n = 3$ degrees of freedom. Consequently the d.f. of the absolute velocity r is

$$P\,(r \leq x) = P\left(\frac{r^2}{\sigma^2} \leq \frac{x^2}{\sigma^2}\right) = \int\limits_0^{x^2/\sigma^2} k_3\,(x)\,d\,x.$$

The fr.f. of r is the derivative of the d.f., or

$$\frac{2\,x}{\sigma^2}\,k_3\left(\frac{x^2}{\sigma^2}\right) = \sqrt{\frac{2}{\pi}}\,\frac{x^2}{\sigma^3}\,e^{-\frac{x^2}{2\,\sigma^2}}.$$

8.2. The t distribution. — Let Y and Z be independent variables such that Y is normal $(0, 1)$, while Z has a χ^2 distribution with n degrees of freedom. Z will then always be positive, and we write

$$t = \sqrt{n}\,\frac{Y}{\sqrt{Z}},$$

where the square root is taken with the positive sign. We shall now try to find the distribution of the random variable t defined by this expression.

The joint fr.f. of Y and Z is

$$\frac{1}{\sqrt{2\,\pi}}\,e^{-\frac{y^2}{2}} \cdot \frac{1}{2^{\frac{n}{2}}\,\Gamma\left(\frac{n}{2}\right)}\,z^{\frac{n}{2}-1}\,e^{-\frac{z}{2}},$$

where y may assume any real value, while z is always positive. If we denote the d.f. of t by $S_n\,(x)$, and in the same way as in the preceding section use C as a notation for an unspecified constant depending only on n, we have

$$S_n(x) = P(t \le x) = P\left(Y \le \frac{x}{\sqrt{n}}\sqrt{Z}\right)$$

$$= C \iint\limits_{y \le x\sqrt{z}/\sqrt{n}} z^{\frac{n}{2}-1} e^{-\frac{y^2}{2}-\frac{z}{2}} dy\, dz.$$

For any given $z > 0$ we have here to integrate with respect to y between the limits $-\infty$ and $x\sqrt{z}/\sqrt{n}$. The double integral is absolutely convergent, and we find by means of simple substitutions

$$S_n(x) = C \int_0^\infty z^{\frac{n}{2}-1} e^{-\frac{z}{2}} dz \int_{-\infty}^{x\sqrt{z}/\sqrt{n}} e^{-\frac{y^2}{2}} dy$$

$$= C \int_0^\infty z^{\frac{n-1}{2}} e^{-\frac{z}{2}} dz \int_{-\infty}^x e^{-\frac{u^2 z}{2n}} du$$

$$= C \int_{-\infty}^x du \int_0^\infty z^{\frac{n-1}{2}} e^{-\frac{1}{2}\left(1+\frac{u^2}{n}\right)z} dz$$

$$= C \int_{-\infty}^x \frac{du}{\left(1+\dfrac{u^2}{n}\right)^{\frac{n+1}{2}}}.$$

If the fr.f. of t is denoted by $s_n(x)$, the last expression gives

(8.2.1) $$s_n(x) = S_n'(x) = \frac{C}{\left(1+\dfrac{x^2}{n}\right)^{\frac{n+1}{2}}}.$$

The constant C is here determined by the condition $\int_{-\infty}^\infty s_n(x)\, dx = 1$, and we give without proof (cf. *Mathematical Methods*, p. 238) the result

$$C = \frac{1}{\sqrt{n\pi}} \cdot \frac{\Gamma\left(\dfrac{n+1}{2}\right)}{\Gamma\left(\dfrac{n}{2}\right)}.$$

The distribution defined by the fr.f. $s_n(x)$ or the d.f. $S_n(x)$ is known as the *t distribution* or STUDENT's *distribution*, after the English statistician W. S. GOSSET, who wrote under the pseudonym of STUDENT.

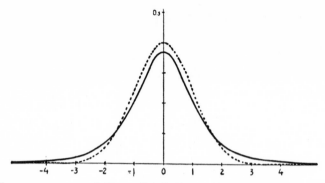

Fig. 21. Frequency curve of the t distribution with $n=3$ degrees of freedom. The dotted curve is a normal frequency curve with $m=0$ and $\sigma=1$.

The parameter n is here, as in the case of the χ^2 distribution, called the number of *degrees of freedom* of the distribution.

The frequency curve $y = s_n(x)$ is symmetric about the point $x = 0$, which is the mean, median and mode of the distribution. We have (cf. also 8.5, ex. 1)

$$E(t) = 0, \quad D^2(t) = \frac{n}{n-2}.$$

For large n the frequency curve runs very close to the normal curve $y = \varphi(x)$. For small n, on the other hand, the t curve deviates considerably from the normal curve in the sense that the probability of a large deviation from the mean is larger according to the t distribution than according to the normal distribution, as may be seen from Fig. 21.

The probability that the random variable t assumes a value such that $|t| > t_0$ is, on account of the symmetry of the distribution (cf. the corresponding relation 7.2.2 for the normal distribution),

$$P(|t| > t_0) = 2(1 - S_n(t_0)).$$

If, conversely, we want to find a quantity t_p such that the probability $P(|t| > t_p)$ takes a given value, say $p \%$, we shall have to solve the equation

$$P(|t| > t_p) = \frac{p}{100}.$$

The unique root t_p of this equation is called the *p percent value of t for n degrees of freedom*. In Table IV this value is given for some values of p and n. It will be seen from this table that for large n the t_p values approximately agree with the λ_p values for the normal distribution. When $n \to \infty$, the t distribution tends to the normal, and t_p tends to λ_p for any given p.

8.3. The F distribution. — If the independent variables U and V have χ^2 distributions with m and n degrees of freedom respectively, the distribution of the variable

$$F = \frac{U/m}{V/n} = \frac{n}{m} \cdot \frac{U}{V}$$

may be found by means of transformations similar to those used in the two preceding sections. A detailed proof is given, e.g., in *Mathematical Methods*, p. 241. Since we have always $F > 0$, the fr.f. of F reduces to zero for $x < 0$, while for $x > 0$ the fr.f. is

$$C \frac{x^{\frac{m}{2}-1}}{(mx+n)^{\frac{m+n}{2}}},$$

where C is a constant depending only on m and n.

The F distribution has very important applications in the statistical method known as the *analysis of variance*, which was introduced by R. A. FISHER. The parameters m and n are called the numbers of degrees of freedom in the numerator and the denominator respectively. The unique root F_p of the equation

$$P(F > F_p) = \frac{p}{100}$$

where p is a given percentage, is called the *p percent value of F for the degrees of freedom m and n*. Some values of F_p are given in Tables V–VI.

8.4. The Pearson system. — The four most important continuous distributions discussed above are the normal and the χ^2, t and F distributions. If $y = f(x)$ denotes the fr.f. of any of these, it is easily verified that y satisfies a differential equation of the form

$$y' = \frac{x+a}{b_0 + b_1 x + b_2 x^2} y,$$

where a, b_0, b_1 and b_2 are constants.

This differential equation forms the base of the system of frequency functions introduced by KARL PEARSON. The form of the solution of the

127

basic differential equation will depend on the properties of the roots of the quadratic polynomial in the denominator of the second member, and accordingly PEARSON distinguished between various types of frequency functions. We shall here only mention some of the most important types.[1] In each case, the constant C should be determined by the condition that the integral of the fr.f., extended over the range of variation given for x, should be equal to unity.

Type I. $\quad y = C\,(x-a)^{p-1}\,(b-x)^{q-1}. \quad a < x < b;\; p > 0,\; q > 0.$

The whole mass of the distribution is here confined between the finite limits a and b. If we take $p = q = \frac{1}{2}\,b^2$, $a = -b$, and allow b to tend to infinity, we obtain the normal distribution as a limiting case.

Type III. $\quad y = C\,(x-\mu)^{\lambda-1}\,e^{-\alpha(x-\mu)}. \quad x > \mu;\; \alpha > 0,\; \lambda > 0.$

For $\mu = 0$, $\lambda = n/2$, $\alpha = \frac{1}{2}$ we obtain the χ^2 distribution as a particular case.

Type VI. $\quad y = C\,(x-a)^{p-1}\,(x-b)^{q-1}. \quad x > b;\; a < b,\; q > 0,\; p+q < 1.$

For $a = -n/m$, $b = 0$, $p = 1 - \frac{1}{2}(m+n)$, $q = \frac{1}{2}m$, we have here the F distribution as a particular case.

A further particular case of type VI is obtained for $a = 0$, $q = 1$, $p = -\alpha$, when the fr.f. becomes $y = C/x^{\alpha+1}$. The probability that the variable assumes a value exceeding x is then of the form C/x^α. This distribution is known as PARETO's distribution and appears often in statistics of incomes and property values. It is an extremely skew distribution, where the values of the mean and the standard deviation are strongly influenced by the small masses situated far out on the positive x-axis, so that it is usually preferable to base the measures of location and dispersion on the median and the quartiles.

Type VII. $\quad y = \dfrac{C}{((x-\alpha)^2 + \beta^2)^m} \cdot \quad -\infty < x < \infty;\; m > \frac{1}{2}.$

It is immediately seen that the t distribution is a particular case of this type.

8.5. Problems. — 1. Find the measures of skewness and excess, γ_1 and γ_2, for the χ^2 and t distributions.

2. Find the mean and the s.d. of the F distribution.

3. Show that the generating function of the χ^2 distribution with n degrees of freedom is $(1-2t)^{-\frac{n}{2}}$. Hence deduce an addition theorem for the χ^2 distribution, similar to those proved in 6.5 and 7.3 for the POISSON and the normal distribution.

---

[1] A complete account of the PEARSON system will be found in ELDERTON's book mentioned in the list of references.

4. Let $x, x_1, \ldots, x_m, y_1, \ldots, y_n$ be independent random variables, all of which are normal $(0, 1)$. Find the fr.f:s of the variables

$$t = \sqrt{m}\,\frac{x}{\sqrt{\sum\limits_{1}^{m} x_i^2}}, \quad u = \sqrt{m}\,\frac{x_1}{\sqrt{\sum\limits_{1}^{m} x_i^2}}, \quad v = \frac{n}{m}\cdot\frac{\sum\limits_{1}^{m} x_i^2}{\sum\limits_{1}^{n} y_j^2}, \quad w = \frac{\sum\limits_{1}^{m} x_i^2}{\sum\limits_{1}^{m} x_i^2 + \sum\limits_{1}^{n} y_j^2}.$$

5. The variable n has a POISSON distribution with a parameter x, which is itself a random variable with the fr.f.

$$f(x) = \frac{\alpha^\lambda}{\Gamma(\lambda)}\, x^{\lambda-1} e^{-\alpha x}, \quad (x > 0, \ \alpha > 0, \ \lambda > 0)$$

of PEARSON's type III. Show that for each $\nu = 0, 1, 2, \ldots$ we have

$$P(n = \nu) = \binom{-\lambda}{\nu} \left(\frac{\alpha}{1+\alpha}\right)^\lambda \left(-\frac{1}{1+\alpha}\right)^\nu.$$

The distribution defined by these probabilities is known as the "negative binomial" or the POLYA distribution.

CHAPTER 9

VARIABLES AND DISTRIBUTIONS IN TWO DIMENSIONS

9.1. Moments and correlation coefficients. — In 5.4 we have introduced some fundamental notions connected with the joint distributions of several random variables. Mean values of functions of several variables have been discussed in 5.5. In the present chapter, we shall give some further developments for the particular case of two variables.

The *moments* of a two-dimensional distribution are defined by a direct generalization of the definition given in 5.6 for the one-dimensional case. When i and k are non-negative integers, we shall thus say that

$$\alpha_{ik} = E(X^i Y^k)$$

is a moment of the joint distribution of the variables X and Y. The sum $i + k$ will be called the order of the moment. In particular the mean values of X and Y are

$$\alpha_{10} = E(X) = m_1, \quad \alpha_{01} = E(Y) = m_2,$$

and these are the coordinates of the centre of gravity of the mass in the two-dimensional distribution. For a distribution belonging to one of the two simple types, we have according to 5.5.4 respectively

$$\alpha_{ik} = \sum_{\mu,\nu} p_{\mu\nu} x_\mu^i y_\nu^k$$

or

$$\alpha_{ik} = \int_{-\infty}^{\infty} \int_{-\infty}^{\infty} x^i y^k f(x, y)\, dx\, dy.$$

The *central moments* of the (X, Y) distribution are the quantities

$$\mu_{ik} = E\left((X - m_1)^i (Y - m_2)^k\right).$$

The central moments of the first order, μ_{10} and μ_{01}, are both equal to zero. The central moments of higher order can be expressed in terms of the ordinary moments α_{ik} by formulae analogous to those given in 5.6 for one-dimensional distributions. Thus we have for the second order central moments

$$
\begin{aligned}
\mu_{20} &= E(X - m_1)^2 = \alpha_{20} - m_1^2 = \sigma_1^2, \\
\mu_{11} &= E(X - m_1)(Y - m_2)) = \alpha_{11} - m_1 m_2, \\
\mu_{02} &= E(Y - m_2)^2 = \alpha_{02} - m_2^2 = \sigma_2^2.
\end{aligned}
$$

(9.1.1)

Thus μ_{20} and μ_{02} are the squares of the standard deviations σ_1 and σ_2. If σ_1 or σ_2 reduces to zero, the corresponding variable has a "one point" distribution (cf. 6.1). In order to avoid trivial complications we assume in the sequel that this case does not arise, so that σ_1 and σ_2 are both positive.

For all real u and v we have

$$(9.1.2) \qquad E(u(X - m_1) + v(Y - m_2))^2 = \mu_{20} u^2 + 2\mu_{11} uv + \mu_{02} v^2 \geq 0,$$

since this is the mean value of a variable which is always non-negative. Taking here in particular $u = \mu_{02}$, $v = -\mu_{11}$, a simple calculation gives

$$(9.1.3) \qquad \mu_{11}^2 \leq \mu_{20}\mu_{02}.$$

We now define the *correlation coefficient* ϱ of the variables X and Y by the expression

$$(9.1.4) \qquad \varrho = \frac{\mu_{11}}{\sqrt{\mu_{20}\mu_{02}}} = \frac{\mu_{11}}{\sigma_1 \sigma_2}.$$

It then follows from 9.1.3 that we have $\varrho^2 \leqq 1$, or

$$-1 \leqq \varrho \leqq 1.$$

Thus the correlation coefficient ϱ always has a value belonging to the interval $(-1, 1)$. In order that ϱ may have one of the extreme values ± 1, the distribution must satisfy certain conditions, and we shall now prove the following important theorem which specifies these conditions.

In order that ϱ may have one of the values ± 1, it is necessary and sufficient that the total mass in the two-dimensional (X, Y) distribution is situated on a straight line in the (X, Y) plane.

In order to show that the condition is necessary, we suppose that $\varrho = \pm 1$. If in 9.1.2 we take $u = \varrho \sigma_2$, $v = -\sigma_1$, the second member reduces to zero. Thus the variable $\varrho \sigma_2 (X - m_1) - \sigma_1 (Y - m_2)$ has its first and second order moments both equal to zero. It follows that the standard deviation is also equal to zero (why?), and consequently this variable has a one point distribution with the total mass concentrated at the origin. This implies, however, that the total mass in the two-dimensional (X, Y) distribution must belong to the line $\varrho \sigma_2 (X - m_1) - \sigma_1 (Y - m_2) = 0$. We note that the angular coefficient of this line is positive when $\varrho = +1$, and negative when $\varrho = -1$.

Suppose, on the other hand, that we know that the total mass in the (X, Y) distribution is situated on some straight line. Then obviously this line must pass through the centre of gravity of the mass, so that its equation can be written in the form $u_0 (X - m_1) + v_0 (Y - m_2) = 0$, where at least one of the coefficients u_0 and v_0 must be different from zero. Suppose, e.g., $u_0 \neq 0$. If in 9.1.2 we take $u = u_0$ and $v = v_0$, the first member becomes the mean value of a "variable" which is always equal to zero, and this is itself equal to zero. Consequently

$$\sigma_1^2 u_0^2 + 2 \varrho \, \sigma_1 \sigma_2 u_0 v_0 + \sigma_2^2 v_0^2 = (1 - \varrho^2) \, \sigma_1^2 u_0^2 + (\varrho \, \sigma_1 u_0 + \sigma_2 v_0)^2 = 0.$$

Since both σ_1 and u_0 are by hypothesis different from zero, this is only possible if $\varrho^2 = 1$, or $\varrho = \pm 1$. Thus the condition is also sufficient, and our theorem is proved.

We have thus seen that, in the extreme cases when $\varrho = \pm 1$, the total mass of the distribution is concentrated on a straight line, and there exists a *complete linear dependence* between the variables X and Y. In fact, if the value assumed by one of the variables is given, there is only one possible value for the other variable, which may be found from the equation of the straight line. Thus either variable is a linear function of the other, and it follows from a remark made above that the variables vary in the same sense, or in inverse senses, according as $\varrho = +1$ or $\varrho = -1$.

On the other hand, we have also seen that such a complete linear dependence will exist *only* when $\varrho = \pm 1$. Thus when ϱ has some intermediate value, there will be no straight line in the (X, Y) plane containing the total mass of the distribution. However, it seems natural to expect that, in the same measure as ϱ approaches one of the extreme values ± 1, the mass will tend to accumulate in the neighbourhood of some straight line. In such a case, the variables would tend to vary in the same sense, or in inverse senses, according as ϱ approaches $+1$ or -1. The correlation coefficient ϱ could then serve as a measure of a property of the distribution that might be called the *degree of linear dependence* between the variables.

We shall return to this subject in the following section. Meanwhile, let us see what happens in the case when there is no dependence whatever between the variables, i.e., when the variables are *independent* according to the definition given in 5.4.

If X and Y are independent, we have by 5.5.8

$$\mu_{11} = E\left((X - m_1)(Y - m_2)\right) = E\left(X - m_1\right) E\left(Y - m_2\right) = 0,$$

so that $\varrho = 0$. When $\varrho = 0$ we shall say that X and Y are *uncorrelated*, and thus we have just proved that *two independent variables are always uncorrelated*.

It should be well observed that the converse statement is *not* true. In other words: *two uncorrelated variables are not necessarily independent*. In order to show this it will be sufficient to give an example of two uncorrelated variables that are not independent. Consider two variables X and Y such that the mass in the two-dimensional joint distribution is spread with constant density over the area of the circle $X^2 + Y^2 \leq 1$. By the symmetry of the distribution, it is then obvious that $\mu_{11} = 0$ and consequently $\varrho = 0$, so that the variables are uncorrelated. On the other hand, it is equally obvious that the variables are not independent. Consider, e.g., the relation

$$P\left(X > V\tfrac{1}{2},\ Y > V\tfrac{1}{2}\right) = P\left(X > V\tfrac{1}{2}\right) \cdot P\left(Y > V\tfrac{1}{2}\right),$$

which will certainly hold true if the variables are independent. For the distribution with which we are now concerned, it is easily seen that the first member of this relation is zero, while the second member is positive, and thus the two members are not equal, so that the variables are not independent. (The reader is recommended to draw the figure, and to find the value of the second member.)

Ex. — For the distribution of 5.4, ex. 1, simple calculations will show that $m_1 = = m_2 = \frac{1}{2}$, $\sigma_1 = \sigma_2 = \frac{1}{2}$, and $\alpha_{11} = \frac{1}{2} p$. By 9.1.1 and 9.1.4 we then have $\mu_{11} = \frac{1}{2} p - \frac{1}{4}$ and $\varrho = 2 p - 1$. For $p = 1$ we thus have $\varrho = 1$, while for $p = 0$ we have $\varrho = -1$. Accordingly we find in both these cases that the total mass of the distribution is situated on a single straight line, viz. on the line $Y = X$ in the case $\varrho = 1$, and on the line $Y = -X + 1$ in the case $\varrho = -1$. — Further, as soon as $p \neq \frac{1}{2}$, we have $\varrho \neq 0$, but for $p = \frac{1}{2}$ we have $\varrho = 0$, which agrees with the fact that the variables are then independent, as we have seen in 5.4.

9.2. Least squares regression.

— As we have seen in 5.4, the joint distribution of two random variables X and Y may be represented by a distribution of mass in the (X, Y) plane. *We shall now try to determine a straight line in the plane by the condition that the line should give the best possible fit to the mass in the distribution.* This is a vaguely expressed condition, and we shall have to make it more precise before we can proceed to the solution of the problem.

Consider a mass particle situated at the point (X, Y). The distance, measured in the direction of the axis of Y, between this particle and the straight line $y = \alpha + \beta x$, is $|Y - \alpha - \beta X|$. It might thus seem natural to try to determine the constants α and β in the equation of the line from the condition that the mean value of the distance, $E(|Y - \alpha - \beta X|)$, should be made as small as possible.

However, it appears here, just as in the case of the choice between the standard deviation and the average deviation as measure of dispersion (cf. 5.7), that we shall arrive at considerably simpler and more manageable results by operating with mean values of *squares*, rather than of *absolute values*. This is but a particular case of a very general principle known as the *least squares principle*, which may be expressed in the following way. Whenever we have to determine some unknown elements (such as constants, functions, curves, etc.) by the condition that certain deviations should be as small as possible, it will generally be advantageous to interpret the condition so as to render the *mean value of the square of the deviation* as small as possible. This is an essentially practical principle, and its main justification lies in the fact that it often leads to simpler calculations and results than any other possible procedure. In the sequel, we shall often have occasion to apply this principle.

In the present case, the principle of least squares leads us to choose as our straight line of closest fit the line that renders the mean value of the square of the distance

$$E\,(Y - \alpha - \beta\,X)^2 = E\,(Y - m_2 - \beta\,(X - m_1) + m_2 - \alpha - \beta\,m_1)^2$$

(9.2.1)

$$= \mu_{02} - 2\,\mu_{11}\,\beta + \mu_{20}\,\beta^2 + (m_2 - \alpha - \beta\,m_1)^2$$

as small as possible. A simple calculation shows that the solution of this minimum problem is given by the expressions

$$\beta = \frac{\mu_{11}}{\mu_{20}} = \frac{\varrho\,\sigma_2}{\sigma_1}, \quad \alpha = m_2 - \beta\,m_1.$$

Thus the straight line $y = \alpha + \beta\,x$, where the constants are given by the above expressions, renders the mean value of the square of the distance, *measured in the direction of the axis of* Y, as small as possible. The equation of this line can be written under any of the equivalent forms

(9.2.2)

$$y = m_2 + \frac{\varrho\,\sigma_2}{\sigma_1}\,(x - m_1),$$

$$\frac{y - m_2}{\sigma_2} = \varrho\,\frac{x - m_1}{\sigma_1}.$$

It will be seen that the line passes through the centre of gravity (m_1, m_2) of the mass, and that its angular coefficient has the same sign as the correlation coefficient ϱ. The expression 9.2.1 of the mean value of the squared distance becomes after introduction of the values of α and β for the minimizing line

(9.2.3)
$$E_{\min}\,(Y - \alpha - \beta\,X)^2 = \frac{\mu_{20}\,\mu_{02} - \mu_{11}^2}{\mu_{20}} = \sigma_2^2\,(1 - \varrho^2).$$

Just as well as we have here measured distances in the direction of the axis of Y, we could measure them in some other way, e.g. in the direction of the axis of X. We shall then only have to exchange X and Y in the preceding argument, and the equation of the minimizing line can now be written under any of the forms

(9.2.4)

$$x = m_1 + \frac{\varrho\,\sigma_1}{\sigma_2}\,(y - m_2),$$

$$\frac{y - m_2}{\sigma_2} = \frac{1}{\varrho}\cdot\frac{x - m_1}{\sigma_1}.$$

Obviously this is also a line passing through the centre of gravity, and with an angular coefficient having the same sign as ϱ, but the two lines 9.2.2 and 9.2.4 are not identical, except in the particular case when ϱ

takes one of the extreme values ± 1. For the line 9.2.4, the mean value of the squared distance is

(9.2.5) $$E_{\min} (X - \alpha - \beta\, Y)^2 = \sigma_1^2 (1 - \varrho^2).$$

The lines 9.2.2 and 9.2.4 are known as the *least squares regression lines* of the distribution. The line 9.2.2 is said to give the *regression of Y on X*, and conversely for 9.2.4. (Cf. Fig. 23, p. 143.) For the coefficient of x in the first equation 9.2.2, and the coefficient of y in the first equation 9.2.4, we shall use the notation

(9.2.6) $$\beta_{21} = \frac{\varrho\, \sigma_2}{\sigma_1}, \qquad \beta_{12} = \frac{\varrho\, \sigma_1}{\sigma_2},$$

and we shall call β_{21} the *regression coefficient of Y on X*, and conversely for β_{12}.

We have thus found that the two regression lines are the straight lines giving the best fit to the mass in the distribution when distances are measured in the direction of the axis of Y for 9.2.2, and the axis of X for 9.2.4. If the standard deviations σ_1 and σ_2 are given, it follows from 9.2.3 and 9.2.5 that for both regression lines the mean value of the squared distance is proportional to $1 - \varrho^2$. The nearer ϱ^2 lies to its maximum value 1, the more the mass in the distribution will thus approach a rectilinear configuration. This fact gives additional support to our tentative conclusion in the preceding section that ϱ may be regarded as a measure of the degree of linear dependence between the variables X and Y.

There is, of course, no necessity for measuring distances in the direction of one of the coordinate axes. We could also measure them, e.g., orthogonally to the straight line that we are dealing with, and try to determine the line so that the mean value of the orthogonal distance becomes as small as possible. The minimizing line obtained in this way is called the *orthogonal regression line* of the distribution. However, the properties of this line are more complicated and less important for most practical applications than those of 9.2.2 and 9.2.4, and will not be discussed here.

Instead of a straight line we may consider any other type of curve, and try to determine the constants in its equation so that the curve will give the best possible fit to the distribution, in the sense of the least squares principle. Consider e.g. the n:th degree parabola with the equation $y = a_0 + a_1 x + \cdots + a_n x^n$. If the coefficients are determined by the condition that the mean value

$$E (Y - a_0 - a_1 X - \cdots - a_n X^n)^2$$

135

should be made as small as possible, we obtain a curve known as the *n:th order regression parabola of Y on X*. It is seen that the coefficients a_ν will be obtained as solutions of a system of linear equations, and that in general there will be one and only one system of solutions. In a similar way we may consider exponential and logarithmic regression curves, etc. In practical applications, the straight regression lines play the most important part, although cases of parabolic, exponential and logarithmic regression are by no means uncommon. — In the following section we shall consider another type of regression curve, which also has important practical applications.

9.3. Conditional distributions. The regression of the mean. — In
this section we shall confine ourselves to the case of continuous distributions, leaving to the reader the task of developing the corresponding formulae for discrete distributions, which will only require simple modifications of the deductions given below.

Let X and Y be two random variables with the joint fr.f. $f(x, y)$. The fr.f:s of the marginal distributions of X and Y are then by 5.4.4 and 5.4.5

$$f_1(x) = \int_{-\infty}^{\infty} f(x, y)\, dy,$$

$$f_2(y) = \int_{-\infty}^{\infty} f(x, y)\, dx.$$

Let us now consider one infinitesimal line element on each coordinate axis, say $(x, x+dx)$ and $(y, y+dy)$. The conditional probability that Y takes a value between y and $y+dy$, relative to the hypothesis that X takes a value between x and $x+dx$, is according to 3.4.1

$$P(y < Y < y+dy \,|\, x < X < x+dx) =$$

$$= \frac{P(x < X < x+dx,\ y < Y < y+dy)}{P(x < X < x+dx)}$$

$$= \frac{f(x, y)\, dx\, dy}{f_1(x)\, dx} = \frac{f(x, y)}{f_1(x)}\, dy.$$

For the coefficient of dy in the last expression we shall introduce the notation

(9.3.1)
$$f(y\,|\,x) = \frac{f(x, y)}{f_1(x)}.$$

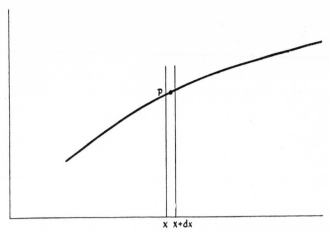

x x+dx

Fig. 22. The regression curve of the Y means. P is the centre of gravity for the mass in the vertical strip between x and $x + dx$.

For every fixed x such that $f_1(x) \neq 0$, the function $f(y\,|\,x)$ has the characteristic properties of a fr. f. in the variable y. In fact, we have always $f(y\,|\,x) \geq 0$, and further

$$\int_{-\infty}^{\infty} f(y\,|\,x)\,dy = \frac{\int_{-\infty}^{\infty} f(x,y)\,dy}{f_1(x)} = 1.$$

We shall call $f(y\,|\,x)$ the *conditional frequency function of Y, relative to the hypothesis that X takes the value x*. The distribution defined by this frequency function is the *conditional distribution of Y*, relative to the same hypothesis.

The *conditional mean value* of Y, relative to the hypothesis $X = x$, is

(9.3.2) $$E(Y\,|\,X = x) = \int_{-\infty}^{\infty} y\,f(y\,|\,x)\,dy = \frac{\int_{-\infty}^{\infty} y\,f(x,y)\,dy}{\int_{-\infty}^{\infty} f(x,y)\,dy}.$$

The point with the coordinates x and $E(Y\,|\,X = x)$ will evidently be the centre of gravity of the mass in the infinitely narrow strip bounded by the verticals through x and $x + dx$. When x varies, the equation

(9.3.3) $$y = E(Y\,|\,X = x)$$

represents a curve which is the locus of all these centres of gravity (cf. Fig. 22).

137

This curve is known as the *regression curve for the mean of Y*. It is, in fact, easily seen that the curve is a regression curve in the sense of the least squares principle, as defined in the preceding section. In order to show this, let us try to find, *among all possible curves* $y = g(x)$, the one that renders the mean square distance $E(Y - g(X))^2$ as small as possible. According to 9.3.1 we have

(9.3.4)

$$E(Y - g(X))^2 = \int_{-\infty}^{\infty} \int_{-\infty}^{\infty} (y - g(x))^2 f(x, y) \, dx \, dy$$

$$= \int_{-\infty}^{\infty} f_1(x) \, dx \int_{-\infty}^{\infty} (y - g(x))^2 f(y \mid x) \, dy.$$

Here, the integral with respect to y is the second order moment of the conditional Y distribution about the point $g(x)$, and we have seen in 5.6 that for any distribution this moment will assume its minimum value when $g(x)$ coincides with the mean of the distribution. Hence the mean value 9.3.4 will become a minimum if for every x we choose $g(x)$ equal to the mean of the conditional Y distribution corresponding to x. Evidently the regression curve $y = g(x)$ will then be identical with 9.3.3.

In the particular case when the regression curve of the mean of Y is a straight line, it can easily be shown that this line is identical with the least squares regression line of Y on X, the equation of which we have given in 9.2.2. In fact, the regression curve of the mean is, *among all possible curves*, the one that renders the mean square distance 9.3.4 as small as possible. If that curve is a straight line, it must obviously coincide with the line that gives, *among all straight lines*, the smallest possible mean square distance. — When this property holds for a given distribution, we shall say that we are concerned with a case of *linear regression*. In the following section, we shall study an important distribution where this case occurs.

By permutation of the variables X and Y in the preceding argument we obtain the conditional fr.f. of X, relative to the hypothesis that Y takes a given value y:

$$f(x \mid y) = \frac{f(x, y)}{f_2(y)}.$$

The corresponding conditional mean value of X generates the *regression curve for the mean of X:*

$$x = E(X \mid Y = y).$$

138

In the particular case when this curve is a straight line, it coincides with the least squares regression line 9.2.4 of X on Y.

Consider now the particular case when X and Y are independent. It then follows from 5.4.9 that we have $f(x, y) = f_1(x) f_2(y)$, and consequently

$$f(y \mid x) = f_2(y), \qquad f(x \mid y) = f_1(x).$$

The first relation shows that the conditional Y distribution is independent of X, so that the regression curve of the mean of Y must be a straight line parallel to the axis of X. In the same way it is seen that the regression curve for the mean of X is a straight line parallel to the axis of Y. These results agree with 9.2.2 and 9.2.4, since for independent variables we have $\varrho = 0$.

Ex. — It is useful to observe that, according to 9.2.6, we have $\varrho^2 = \beta_{21} \beta_{12}$, where all three quantities ϱ, β_{21} and β_{12} have the same sign. From this relation we can find the correlation coefficient ϱ, when the two regression coefficients β_{21} and β_{12} are known.

As an example of the use of this remark, we consider the continuous distribution where the mass is spread with constant density over the area of the parallelogram bounded by the lines $y = 0$, $y = 2$, $y = x$ and $y = x - 2$. It will be seen without difficulty (draw the figure!) that all conditional distributions of Y as well as of X are rectangular distributions, and that the locus of the conditional Y means is the diagonal $y = \frac{1}{2} x$, while the locus of the conditional X means is the line $y = x - 1$. Thus both regression curves of the means are straight lines, and as such they must coincide with the least squares regression lines 9.2.2 and 9.2.4. It follows that we have $\beta_{21} = \frac{1}{2}$ and $\beta_{12} = 1$. Hence we obtain $\varrho^2 = \frac{1}{2}$ and $\varrho = \sqrt{\frac{1}{2}}$, since ϱ has the same sign as β_{21} and β_{12}.

When the regression of both variables is linear, i.e , when the regression curves of the X and Y means are both straight lines, it follows from 9.2.3 and 9.2.5 that the correlation coefficient ϱ gives a measure of the degree of concentration of the probability mass about any of these lines. If the regression curves are at least approximately straight lines, this will still be approximately true. But in a case when regression is not even approximately linear, it may be of greater interest to consider a measure of the *concentration of the mass about one of the (non-linear) regression curves of the means*. Let us denote the second member of 9.3.3 by $g(x)$, so that $y = g(x)$ is the equation of the regression curve of the Y means. We then have, using 9.3.1,

$$E(Y - g(X))(g(X) - m_2) = \int_{-\infty}^{\infty} \int_{-\infty}^{\infty} (y - g(x))(g(x) - m_2) f(x, y) \, dx \, dy$$

$$= \int_{-\infty}^{\infty} (g(x) - m_2) f_1(x) \, dx \int_{-\infty}^{\infty} (y - g(x)) f(y \mid x) \, dy = 0,$$

since $g(x)$ is the conditional mean of Y. Hence we obtain

$$\sigma_2^2 = E\,(Y - m_2)^2$$
$$= E\,(Y - g\,(X) + g\,(X) - m_2)^2$$
$$= E\,(Y - g\,(X))^2 + E\,(g\,(X) - m_2)^2.$$

Thus if we define a non-negative quantity $\eta = \eta_{YX}$ by the relation

(9.3.5) $$\eta^2 = \frac{1}{\sigma_2^2}\,E\,(g\,(X) - m_2)^2,$$

we have

(9.3.6) $$\eta^2 = 1 - \frac{1}{\sigma_2^2}\,E\,(Y - g\,(X))^2.$$

From the two last relations, it follows that we have always $0 \le \eta^2 \le 1$. Moreover, it follows from 9.3.6 that we have $\eta^2 = 1$ if and only if the total mass of the distribution is concentrated on the curve $y = g(x)$, while 9.3.5 shows that $\eta^2 = 0$ if and only if $g(x)$ reduces to the constant m_2, i.e., when the regression curve for the mean of Y is a horizontal straight line. The quantity $\eta = \eta_{YX}$ is known as the *correlation ratio of Y on X*, and may be regarded as a measure of the concentration of the mass in the distribution about the regression curve of the Y means. — By permutation of the variables X and Y we define in a similar way the correlation ratio η_{XY} of X on Y.

9.4. The normal distribution. — The one-dimensional normal distribution has been introduced in 7.1 and 7.2. Before attempting to generalize this to two dimensions, we recall the following simple facts implied by our discussion in 7.2. — If a variable X' has the standardized normal fr.f.

$$\varphi\,(x') = \frac{1}{\sqrt{2\pi}}\,e^{-\frac{x'^2}{2}},$$

any linear function $X = m + \sigma X'$ is said to be normally distributed. Assuming $\sigma > 0$, the fr.f. of X is

$$\frac{1}{\sigma}\,\varphi\left(\frac{x - m}{\sigma}\right) = \frac{1}{\sigma\sqrt{2\pi}}\,e^{-\frac{(x - m)^2}{2\sigma^2}},$$

and *this is the general form of the one-dimensional normal distribution.*

We now proceed to introduce the *two-dimensional normal distribution* by an appropriate generalization of this procedure. Consider first two independent variables X' and Y', each of which is normal $(0, 1)$. By 5.4.9 the joint fr.f. of X' and Y' is

(9.4.1) $$\varphi(x')\,\varphi(y') = \frac{1}{2\pi}\,e^{-\frac{1}{2}(x'^2+y'^2)}.$$

Now, let m_1 and m_2 denote two real constants, while σ_1 and σ_2 are two positive constants, and ϱ a constant satisfying the inequality $-1 < \varrho < 1$. We then introduce new variables X and Y by the substitution

$$X = m_1 + \sigma_1 X',$$
$$Y = m_2 + \varrho\,\sigma_2 X' + \sqrt{1-\varrho^2}\,\sigma_2 Y'.$$

The substitution has been so arranged that the new variables have the means m_1 and m_2, the standard deviations σ_1 and σ_2, and the correlation coefficient ϱ. This is easily verified by means of 5.5.7, 5.7.5, and 9.1.4. Moreover, it follows from 7.3 that each of the new variables is normally distributed. Summing up, we find that X is normal (m_1, σ_1), while Y is normal (m_2, σ_2), and the correlation coefficient of X and Y is ϱ. — Solving with respect to X' and Y', we obtain

(9.4.2)
$$X' = \frac{X - m_1}{\sigma_1},$$
$$Y' = \frac{1}{\sqrt{1-\varrho^2}}\left(-\varrho\,\frac{X - m_1}{\sigma_1} + \frac{Y - m_2}{\sigma_2}\right).$$

According to 5.4.6, the fr.f. of the new variables X and Y is obtained by introducing the substitution 9.4.2 into 9.4.1, and multiplying by the absolute value of the determinant of the substitution. A simple calculation then leads to the fr.f.

(9.4.3) $$f(x, y) = \frac{1}{2\pi\,\sigma_1\sigma_2\,\sqrt{1-\varrho^2}}\,e^{-\frac{1}{2}Q(x,y)},$$

where

(9.4.4) $$Q(x, y) = \frac{1}{1-\varrho^2}\left(\frac{(x-m_1)^2}{\sigma_1^2} - \frac{2\varrho\,(x-m_1)\,(y-m_2)}{\sigma_1\sigma_2} + \frac{(y-m_2)^2}{\sigma_2^2}\right).$$

This is the general form of the two-dimensional normal frequency function.

The *marginal distributions* (cf. 5.4) of X and Y are already known to be normal (m_1, σ_1) and (m_2, σ_2) respectively. The corresponding fr.f:s are

$$f_1(x) = \frac{1}{\sigma_1 \sqrt{2\pi}} e^{-\frac{(x-m_1)^2}{2\sigma_1^2}},$$

$$f_2(y) = \frac{1}{\sigma_2 \sqrt{2\pi}} e^{-\frac{(y-m_2)^2}{2\sigma_2^2}}.$$

The *conditional fr.f.* of Y, relative to the hypothesis that X takes a given value x, is obtained from 9.3.1, and we find after some calculation

$$f(y|x) = \frac{f(x,y)}{f_1(x)} = \frac{1}{\sigma_2 \sqrt{1-\varrho^2} \cdot \sqrt{2\pi}} e^{-\frac{\left(y-m_2-\frac{\varrho\sigma_2}{\sigma_1}(x-m_1)\right)^2}{2\sigma_2^2(1-\varrho^2)}}.$$

This is a normal fr.f. in y, with the mean value

$$E(Y|X=x) = m_2 + \frac{\varrho\sigma_2}{\sigma_1}(x-m_1),$$

which is a linear function of the given value of X. Thus for the normal distribution the regression curve 9.3.3 for the mean of Y is a straight line, and in accordance with the argument of 9.3 we find that this line is identical with the least squares regression line 9.2.2 of Y on X. — We further note that the s.d. of the conditional Y distribution, $\sigma_2 \sqrt{1-\varrho^2}$, is independent of the given value of X.

By interchanging the variables X and Y in the preceding argument we obtain the conditional fr.f. of X, relative to the hypothesis that Y takes a given value y. This is again a normal fr.f., the mean of which is a linear function of y, so that the regression curve for the mean of X is also a straight line, with the equation 9.2.4. The s.d. of the conditional X distribution is $\sigma_1 \sqrt{1-\varrho^2}$, which is independent of the given value of Y.

We have thus seen that for the normal distribution the regression curves for the means of both variables are straight lines, so that the normal distribution constitutes a case of *linear regression* in respect of both variables.

Consider now the *normal frequency surface*

$$z = f(x,y),$$

where the axis of z stands vertically on a horizontal (x,y) plane, and $f(x,y)$ is given by 9.4.3. The intersection between this surface and a plane parallel to the (y,z) plane, say the plane $x=x_0$, is given by the equation

$$z = f(x_0, y) = f_1(x_0) f(y|x_0).$$

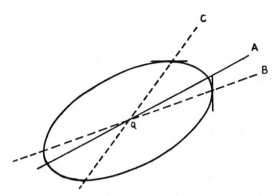

Fig. 23. Equiprobability ellipse and regression lines. Q is the centre of gravity of the distribution, QA is the major axis of the ellipse, QB and QC are the least squares regression lines of Y on X, and of X on Y, respectively. Note the directions of the tangents in the points of intersection of the ellipse with QB and QC.

Disregarding the constant factor $f_1(x_0)$, this is a normal frequency curve, the mean of which is situated on the straight regression line of the Y means. — The intersections with planes parallel to the (x, z) plane have similar properties.

On the other hand, for the intersection between the frequency surface and a plane parallel to the (x, y) plane, say $z = z_0$, we have the equation

$$Q(x, y) = -2 \log \left(2 \pi \sigma_1 \sigma_2 z_0 \sqrt{1 - \varrho^2}\right).$$

If this equation is taken alone, it obviously represents the projection in the (x, y) plane of the curve of intersection. For $0 < z_0 < \dfrac{1}{2 \pi \sigma_1 \sigma_2 \sqrt{1 - \varrho^2}}$ the second member of the equation is positive, and the equation represents an ellipse (prove this!) in the (x, y) plane, with its centre at the point (m_1, m_2). When z_0 varies, we obtain a family of concentric and homothetic ellipses. For $z_0 = \dfrac{1}{2 \pi \sigma_1 \sigma_2 \sqrt{1 - \varrho^2}}$ the ellipse reduces to the point (m_1, m_2), while for $z_0 = 0$ it becomes infinitely large. Each of these ellipses is a curve of constant height for the surface $z = f(x, y)$, and consequently a *curve of constant probability density*, or an *equiprobability curve*, of the normal distribution.

The normal frequency surface $z = f(x, y)$ reaches its maximum height over the (x, y) plane at the common centre (m_1, m_2) of the ellipses, where we have $z = \dfrac{1}{2 \pi \sigma_1 \sigma_2 \sqrt{1 - \varrho^2}}$. For all points of a given equiprobability

143

ellipse, the surface has a constant height, which decreases with increasing dimensions of the ellipse. The form of the ellipses thus gives a good idea of the shape of the frequency surface. If for simplicity we assume $\sigma_1 = \sigma_2$, it can be shown without difficulty by means of the expression 9.4.4 of $Q(x, y)$ that all the ellipses have the common excentricity $\dfrac{2|\varrho|}{1+|\varrho|}$. It follows that when $\varrho = 0$ the ellipses are circles (this is, of course, seen without any calculation from 9.4.4), while as ϱ approaches $+1$ or -1 they become oblong and narrow, thus indicating a tendency of the mass in the distribution to approach a rectilinear configuration (cf. 9.2).

Further, it can be shown that the common major axis of the equiprobability ellipses is the orthogonal regression line (cf. 9.2) of the distribution, while the ordinary regression line of the Y means is the locus of those points of the ellipses where the tangent is parallel to the Y axis, and similarly for the regression line of the X means (cf. Fig. 23).

If in the normal fr.f. 9.4.3 we introduce new variables by a linear substitution of non-vanishing determinant, this will only imply a change of coordinates for the equiprobability ellipses $Q(x, y) = $ const., while the characteristic properties of the distribution will remain invariant. We thus have the following theorem: *If X and Y have a normal joint frequency function, the same will be true of any two linear functions of X and Y, provided that the determinant of the two functions is different from zero.*

When the correlation coefficient ϱ is equal to zero, it is immediately seen from 9.4.3 and 9.4.4 that the normal fr.f. factors in the form $f(x, y) = f_1(x) f_2(y)$. Then by 5.4.9 the variables X and Y are independent. *Thus two normally distributed variables are independent when and only when they are uncorrelated.* — It is important to note that, according to 9.1, this property does *not* in general hold for other than normally distributed variables.

9.5. Problems. — 1. Show that $E(XY) = E(X)E(Y)$ and $D^2(X+Y) = = D^2(X) + D^2(Y)$ when and only when X and Y are uncorrelated (cf. 5.5).

2. X is randomly chosen between -1 and $+1$. Discuss the joint distributions of the following pairs of variables: a) X and X^2, b) X and X^3, c) X and $\sin \frac{1}{2}\pi X$, d) $\sin \frac{1}{2}\pi X$ and $\cos \frac{1}{2}\pi X$. Find in each case the curve on which the total mass of the distribution is situated. Find the least squares regression lines and the correlation coefficient.

3. X and Y have the s.d:s σ_1 and σ_2, and the correlation coefficient ϱ. Find the angle between the two least squares regression lines.

4. Find the condition that must be satisfied by the constants a, b, c and d in order that the variables $U = aX + bY$ and $V = cX + dY$ should be uncorrelated.

5. A human population is composed of two races R_1 and R_2, which are represented in the proportions p and $q = 1 - p$ respectively. For individuals belonging to R_1, the anthropometric characteristics X and Y are assumed to be uncorrelated variables with means m_1 and n_1, and a common standard deviation s. The same holds true for individuals belonging to R_2, although the mean values are here m_2 and n_2, while the s.d. has the same value s. Find the correlation coefficient ϱ between the characteristics X and Y for an individual randomly selected from the total population. Study the behaviour of ϱ when s tends to zero.

6. $X_1, \ldots, X_n$ are uncorrelated variables, all with the same distribution. The arithmetic mean of the variables is denoted by $\overline{X}$. Find, for arbitrary i, the correlation coefficient between a) X_i and $\overline{X}$, b) $X_i - \overline{X}$ and $\overline{X}$.

7. Two non-negative random variables have the joint fr.f. $f(x, y) = \frac{1}{2} x^3 e^{-x(y+1)}$ ($x > 0$, $y > 0$). Find the correlation coefficient, the least squares regression lines, the regression curves for the means, and the correlation ratios.

8. Two random variables have the least squares regression lines $3x + 2y - 26 = 0$ and $6x + y - 31 = 0$. Find the mean values and the correlation coefficient.

9. X and Y have a normal joint distribution. Find the probability that the point (X, Y) is situated outside the ellipse

$$\frac{1}{2(1-\varrho^2)} \left(\frac{(x-m_1)^2}{\sigma_1^2} - \frac{2\varrho(x-m_1)(y-m_2)}{\sigma_1 \sigma_2} + \frac{(y-m_2)^2}{\sigma_2^2} \right) = c^2.$$

10. Prove the statement made in 9.4 that the regression line for the Y means of two-dimensional normal distribution is the locus of those points of the equiprobability ellipses where the tangent is parallel to the Y axis, and similarly for the other regression line.

11. The mass in a certain distribution is spread with constant density over the area of the ellipse in problem 9. Find the conditional distribution of each variable, the regression curves for the means, and the correlation coefficient.

CHAPTER 10

VARIABLES AND DISTRIBUTIONS IN MORE THAN TWO DIMENSIONS

10.1. Regression. — A great part of the contents of the preceding chapter can be directly generalized to the case of a joint distribution of more than two variables. This subject will only be very briefly treated here.

Throughout the whole of this chapter we shall make the simplifying assumption that all appearing random variables have the mean value zero. The formulae corresponding to the general case will then be obtained by replacing everywhere X_1 by $X_1 - m_1$, etc.

Let the variables $X_1, \ldots, X_n$ have a given joint distribution, such that X_i has the s.d. σ_i, while X_i and X_k have the correlation coefficient ϱ_{ik}.

145

We shall now try to find that linear combination of $X_2, \ldots, X_n$ which, in the sense of the least squares principle (cf. 9.2) gives the best fit to the variable X_1. We then regard X_1, as it were, as a "dependent" variable, and try to represent it as well possible by a linear combination of the "independent" variables $X_2, \ldots, X_n$. Accordingly we shall have to determine the coefficients β so as to minimize the mean square

$$(10.1.1) \qquad E\,(X_1 - \beta_{12\cdot 34\ldots n}\,X_2 - \cdots - \beta_{1\,n\cdot 23\ldots n-1}\,X_n)^2.$$

Evidently this is a straightforward generalization of the problem solved in 9.2, which led to the introduction of the two least squares regression lines 9.2.2 and 9.2.4. By analogy the generalized "plane" in the n-dimensional space represented by the equation

$$x_1 = \beta_{12\cdot 34\ldots n}\,x_2 + \cdots + \beta_{1\,n\cdot 23\ldots n-1}x_n,$$

where the coefficients β are determined so as to minimize 10.1.1 will be called the *least squares regression plane of* X_1 *on* $X_2, \ldots, X_n$. The β are the corresponding *regression coefficients*, and the notation used above for these coefficients should be so interpreted that, e. g., $\beta_{12\cdot 3\ldots n}$ denotes the regression coefficient of X_1 on X_2 when $X_3, \ldots, X_n$ are used as variables in the regression equation besides X_2. The two first subscripts on each β are called the *primary subscripts*, while the subscripts appearing after the point are the *secondary subscripts*. The latter will often be omitted when no misunderstanding seems likely to arise.

According to the ordinary theory of maxima and minima, the β have to be determined by the condition that the derivative of 10.1.1 with respect to each β should be equal to zero. This gives $n-1$ linear equations for the $n-1$ unknown β. Omitting the secondary subscripts, these equations are

$$\varrho_{22}\,\sigma_2\,\sigma_2\,\beta_{12} + \cdots + \varrho_{2n}\,\sigma_2\,\sigma_n\,\beta_{1n} = \varrho_{21}\,\sigma_2\,\sigma_1,$$

$$\cdot \quad \cdot \quad \cdot \quad \cdot \quad \cdot \quad \cdot \quad \cdot \quad \cdot \quad \cdot \quad \cdot \quad \cdot \quad \cdot \quad \cdot \quad \cdot \quad \cdot \quad \cdot \quad \cdot \quad \cdot \quad \cdot$$

$$\beta_{n2}\,\sigma_n\,\sigma_2\,\beta_{12} + \cdots + \varrho_{nn}\,\sigma_n\,\sigma_n\,\beta_{1n} = \varrho_{n1}\,\sigma_n\,\sigma_1.$$

Introducing the substitution

$$\gamma_i = \frac{\sigma_i}{\sigma_1}\,\beta_{1i} \qquad (i = 2, \ldots, n),$$

we obtain the new system of equations

$$\varrho_{22}\gamma_2 + \varrho_{23}\gamma_3 + \cdots + \varrho_{2n}\gamma_n = \varrho_{21},$$

$$\cdot \quad \cdot \quad \cdot \quad \cdot \quad \cdot \quad \cdot \quad \cdot \quad \cdot \quad \cdot \quad \cdot \quad \cdot \quad \cdot$$

$$\varrho_{n2}\gamma_2 + \varrho_{n3}\gamma_3 + \cdots + \varrho_{nn}\gamma_n = \varrho_{n1}.$$

Now, let P denote the determinant of all correlation coefficients ϱ_{ik}:

$$P = \begin{vmatrix} \varrho_{11} & \varrho_{12} & \cdots & \varrho_{1n} \\ \varrho_{21} & \varrho_{22} & \cdots & \varrho_{2n} \\ \cdot & \cdot & \cdot & \cdot \\ \varrho_{n1} & \varrho_{n2} & \cdots & \varrho_{nn} \end{vmatrix}.$$

This is a symmetric determinant, since $\varrho_{ik} = \varrho_{ki}$. Moreover, the elements in the main diagonal are all equal to 1, since ϱ_{ii} is the correlation coefficient of two identical variables, and thus $\varrho_{ii} = 1$. We further denote by P_{ik} the cofactor of the element ϱ_{ik} in the determinant P. The determinant of the system of equations satisfied by $\gamma_2, \ldots, \gamma_n$ is then P_{11}, and if we assume $P_{11} \neq 0$, it is well known that the system has a unique solution expressed by the relations

$$\gamma_i = -\frac{P_{1i}}{P_{11}} \quad (i = 2, \ldots, n),$$

which give

(10.1.2) $$\beta_{1i} = -\frac{\sigma_1}{\sigma_i} \cdot \frac{P_{1i}}{P_{11}}.$$

Just as in the particular case $n = 2$, the regression coefficients β_{1i} can thus be found as soon as we know the standard deviations σ_i and the correlation coefficients ϱ_{ik}. In the case $n = 2$, the above expressions give $P = \begin{vmatrix} 1 & \varrho \\ \varrho & 1 \end{vmatrix}$, $P_{11} = 1$, $P_{12} = -\varrho$, and thus $\beta_{12} = \frac{\varrho\sigma_1}{\sigma_2}$, in agreement with 9.2.6.

The β_{1i} given by 10.1.2 are the coefficients in the *best linear estimate of* X_1 *in terms of* $X_2, \ldots, X_n$, in the sense explained above. The difference

(10.1.3) $$Y_{1 \cdot 23 \ldots n} = X_1 - (\beta_{12}X_2 + \cdots + \beta_{1n}X_n)$$

represents that part of X_1 that is left after subtraction of this best linear estimate. We shall call $Y_{1 \cdot 23 \ldots n}$ the *residual* of X_1 with respect to $X_2, \ldots, X_n$. The mean of the residual is clearly equal to zero, while the variance, which is denoted as the *residual variance* of X_1, is

(10.1.4) $$\sigma^2_{1 \cdot 23 \ldots n} = E(Y^2_{1 \cdot 23 \ldots n}) = \sigma^2_1 \frac{P}{P_{11}}.$$

147

It is easily verified that for $n=2$ this expression takes the value $\sigma_1^2(1-\varrho_{12}^2)$, in agreement with 9.2.5. For a proof of 10.1.4 for an arbitrary n, as well as for proofs of some of the properties mentioned in the sequel, we refer to *Mathematical Methods*, Ch. 23.

It can be shown that we always have $P \geqq 0$ and $P_{ii} \geqq 0$ for every $i = 1, 2, \ldots, n$.

If all correlation coefficients ϱ_{ik} with $i \neq k$ are equal to zero, the variables $X_1, \ldots, X_n$ are said to be *uncorrelated*. In this case, it is easily seen that $P = P_{ii} = 1$, while $P_{ik} = 0$ for $i \neq k$. It follows from 10.1.2 that all regression coefficients β_{1i} vanish, so that the equation of the regression plane for X_1 reduces to $x_1 = 0$.

By an appropriate permutation of the variables in the preceding formulae we obtain, of course, regression planes and residuals for any given variable with respect to the others.

10.2. Partial correlation. — For the degree of linear dependence or covariation between two variables X_1 and X_2, we have found an adequate measure (cf. 9.1 and 9.2) in the correlation coefficient ϱ_{12}. However, if X_1 and X_2 are considered in conjunction with some other variables, say $X_3, \ldots, X_n$, it is natural to assume that the covariation between X_1 and X_2 may to a certain extent be due to a common influence from the other variables. It would then be desirable to eliminate this influence, and so to find a measure of the remaining degree, if any, of linear covariation between X_1 and X_2.

As we have seen in the preceding section, the residuals $Y_{1 \cdot 34 \ldots n}$ and $Y_{2 \cdot 34 \ldots n}$ represent those parts of X_1 and X_2, respectively, which remain after subtraction of the best linear estimate of each variable in terms of $X_3, \ldots, X_n$. Accordingly we shall adopt the correlation coefficient between these two residuals as our measure of the degree of linear covariation between X_1 and X_2, after removal of any influence due to the variables $X_3, \ldots, X_n$. Since both residuals have the mean zero, the correlation coefficient is

$$\varrho_{12 \cdot 34 \ldots n} = \frac{E(Y_{1 \cdot 34 \ldots n} \, Y_{2 \cdot 34 \ldots n})}{\sigma_{1 \cdot 34 \ldots n} \, \sigma_{2 \cdot 34 \ldots n}}.$$

This quantity will be called the *partial correlation coefficient* of X_1 and X_2, with respect to $X_3, \ldots, X_n$, as opposed to ϱ_{12}, which is sometimes called the *total correlation coefficient* of X_1 and X_2.

We shall first deduce an expression for the partial correlation coefficient in the particular case $n = 3$, and then give some formulae for the case of a general n, for the proofs of which we refer to *Mathematical Methods*, Ch. 23. By 10.1.2 and 10.1.3 we have for $n = 3$

$$Y_{1 \cdot 3} = X_1 - \frac{\varrho_{13} \sigma_1}{\sigma_3} X_3, \quad Y_{2 \cdot 3} = X_2 - \frac{\varrho_{23} \sigma_2}{\sigma_3} X_3,$$

and hence

$$E(Y_{1 \cdot 3}^2) = \sigma_1^2 (1 - \varrho_{13}^2), \quad E(Y_{2 \cdot 3}^2) = \sigma_2^2 (1 - \varrho_{23}^2),$$

$$E(Y_{1 \cdot 3} Y_{2 \cdot 3}) = \sigma_1 \sigma_2 (\varrho_{12} - \varrho_{13} \varrho_{23}),$$

and

(10.2.1)
$$\varrho_{12 \cdot 3} = \frac{\varrho_{12} - \varrho_{13} \varrho_{23}}{\sqrt{(1 - \varrho_{13}^2)(1 - \varrho_{23}^2)}}.$$

For arbitrary n, there is an analogous formula, viz.,

(10.2.2)
$$\varrho_{12 \cdot 34 \ldots n} = \frac{\varrho_{12 \cdot 34 \ldots n-1} - \varrho_{1 n \cdot 34 \ldots n-1} \varrho_{2 n \cdot 34 \ldots n-1}}{\sqrt{(1 - \varrho_{1 n \cdot 34 \ldots n-1}^2)(1 - \varrho_{2 n \cdot 34 \ldots n-1}^2)}}.$$

In 10.2.1 and 10.2.2 the variables may, of course, be arbitrarily permuted. Thus if we know all total correlation coefficients ϱ_{ik} ($i, k = 1, \ldots, n$), we can in the first place compute all partial correlation coefficients of the form $\varrho_{ij \cdot k}$ by means of 10.2.1 and the corresponding formulae obtained by permutation, and then by means of 10.2.2 proceed to coefficients of the form $\varrho_{ij \cdot kl}$, etc. When all total and partial correlation coefficients are known, we can further compute all residual variances and regression coefficients by means of the relations

$$\sigma_{1 \cdot 23 \ldots n}^2 = \sigma_1^2 (1 - \varrho_{12}^2)(1 - \varrho_{13 \cdot 2}^2)(1 - \varrho_{14 \cdot 23}^2) \ldots (1 - \varrho_{1 n \cdot 23 \ldots n-1}^2),$$

$$\beta_{12 \cdot 34 \ldots n} = \varrho_{12 \cdot 34 \ldots n} \frac{\sigma_{1 \cdot 34 \ldots n}}{\sigma_{2 \cdot 34 \ldots n}},$$

where again the subscripts may be arbitrarily permuted.

In the particular case when the variables $X_1, \ldots, X_n$ are uncorrelated, it follows from the above relations that all partial correlation coefficients corresponding to a subgroup of the variables are zero. On the other hand, a single total correlation coefficient, say ϱ_{12}, may well be equal to zero, while at the same time some partial correlation coefficient between the same variables, say $\varrho_{12 \cdot 3}$, may be different from zero, as will be readily seen from 10.2.1.

Ex. — Let Z_1, Z_2, and Z_3 be random variables, all with zero mean and unity s.d. Suppose that Z_1 and Z_2 have the total correlation coefficient ϱ, while Z_3 is uncorrelated with Z_1 as well as with Z_2. Put

149

$$X_1 = Z_1 + c Z_3,$$

$$X_2 = Z_2 + c Z_3,$$

$$X_3 = Z_3.$$

By some easy calculation, we then find $\varrho_{12} = \dfrac{c^2 + \varrho}{c^2 + 1}$, and $\varrho_{12\cdot 3} = \varrho$. The total correlation coefficient ϱ_{12} of X_1 and X_2 can thus be made to lie as near to one as we please, by choosing c sufficiently large. For large c, there will thus be a high degree of linear covariation between X_1 and X_2. However, this covariation is obviously due to the common influence of X_3, and the covariation left after removal of this influence has the measure $\varrho_{12\cdot 3} = \varrho$, where ϱ may be given any value whatever between -1 and $+1$.

10.3. Multiple correlation. — Consider the correlation coefficient between the variables X_1 and X_1^*, where

$$X_1^* = \beta_{12} X_2 + \cdots + \beta_{1n} X_n$$

is the best linear estimate of X_1 in terms of the variables $X_2, \ldots, X_n$ This will be denoted by $\varrho_{1(23\ldots n)}$, and called the *multiple correlation coefficient* between X_1 on the one side, and the totality of the variables $X_2, \ldots, X_n$ on the other. It can be shown (*Mathematical Methods*, Ch. 23) that we have

$$(10.3.1) \qquad \varrho_{1(23\ldots n)} = \sqrt{1 - \frac{P}{P_{11}}} = \sqrt{1 - \frac{\sigma_{1\cdot 23\ldots n}^2}{\sigma_1^2}},$$

where the square roots have to be taken with the positive sign. Thus we have always $0 \leq \varrho_{1(23\ldots n)} \leq 1$. When $X_1, \ldots, X_n$ are uncorrelated, we have $\varrho_{1(23\ldots n)} = 0$.

Ex. — Let Z_1, Z_2, and Z_3 be uncorrelated variables, each of which has zero mean and unity s.d. Put

$$X_1 = Z_1 + Z_2,$$

$$X_2 = Z_2 + Z_3,$$

$$X_3 = Z_3 + Z_1.$$

We then find $\varrho_{ik} = \frac{1}{2}$ for $i \neq k$, and hence the determinants $P = \frac{1}{2}$, $P_{11} = \frac{3}{4}$, $P_{12} = P_{13} = -\frac{1}{4}$, and the regression coefficients $\beta_{12\cdot 3} = \beta_{13\cdot 2} = \frac{1}{3}$. Thus the best linear estimate of X_1 in terms of X_2 and X_3 is $X_1^* = \frac{1}{3}(X_2 + X_3) = \frac{1}{3}(Z_1 + Z_2) + \frac{2}{3} Z_3$. The multiple correlation coefficient $\varrho_{1(23)}$, or the correlation coefficient between X_1 and X_1^*, is

$$\varrho_{1(23)} = \sqrt{1 - \frac{P}{P_{11}}} = \frac{1}{\sqrt{3}}.$$

10.4. The normal distribution. — The two-dimensional normal frequency function given by 9.4.3 and 9.4.4 can be generalized to an arbitrary number of dimensions. As in the preceding sections of the present chapter, we shall be concerned with n random variables $X_1, \ldots, X_n$, each of which is assumed to have the mean value zero. The general form of the normal fr.f. with zero means is

$$f(x_1, \ldots x_n) = \frac{1}{(2 n)^{n/2} \sigma_1 \ldots \sigma_n \sqrt{P}} e^{-\frac{1}{2} Q(x_1, \ldots, x_n)}$$

where

$$Q(x_1, \ldots, x_n) = \frac{1}{P} \sum_{i, k} P_{i k} \frac{x_i x_k}{\sigma_i \sigma_k}, \quad (i, k = 1, 2, \ldots, n),$$

where P and $P_{i k}$ denote the determinants introduced in 10.1. When $n = 2$, these expressions are easily seen to reduce to 9.4.3 and 9.4.4.

The equation $Q(x_1, \ldots, x_n) = $ const. represents a generalized "ellipsoid" in the n-dimensional space of the coordinates $x_1, \ldots, x_n$, and for different values of the constant we obtain a family of homothetic ellipsoids, each of which is a locus of points of constant probability density for the normal distribution. These are the *equiprobability ellipsoids* of the distribution.

In the same way as in the two-dimensional case, we obtain the following theorem: *If the variables $X_1, \ldots, X_n$ have a joint normal fr.f., the same will hold true of any system of n linear functions of $X_1, \ldots, X_n$, provided that the determinant of these linear functions is different from zero.*

When $X_1, \ldots, X_n$ are uncorrelated, we have (cf. 10.1) $P = P_{i i} = 1$, and $P_{i k} = 0$ for $i \neq k$. Consequently in this case the normal fr.f. $f(x_1, \ldots, x_n)$ breaks up into a product of one-dimensional fr.f:s, and we have the following generalization of a result given in 9.4: *n normally distributed variables are independent when and only when they are uncorrelated.*

10.5. Problems. — **1.** At a deal in bridge a certain player receives X_1 spades, X_2 hearts, X_3 diamonds, and X_4 clubs. Find the conditional mean of an arbitrary X_i, relative to the hypothesis that another arbitrary X_k takes a given value. Use this result to find the total correlation coefficients ϱ_{ij}, the partial correlation coefficients of the forms $\varrho_{ij \cdot k}$ and $\varrho_{ij \cdot kl}$, and the multiple correlation coefficients of the forms $\varrho_{i(jk)}$ and $\varrho_{i(jkl)}$.

2. At a deal in bridge the player A receives X_1 hearts, while B receives X_2 hearts and X_3 diamonds. Find all total, partial, and multiple correlation coefficients.

3. In a certain joint distribution of three variables we have $\varrho_{12} = \varrho_{13} = \varrho_{23} = c$. What values are possible for the constant c?

4. The variables X, Y, and Z satisfy the identity $a X + b Y + c Z = 0$. Each variable has the mean zero and the s.d. σ. Find the correlation coefficients ϱ_{ij}.

5. Each of the variables X, Y, and Z has the mean zero and the s.d. σ. Find a necessary and sufficient condition that the variables $U = Y + Z$, $V = X + Z$, and $W = X + Y$ are uncorrelated.

PART III

APPLICATIONS

CHAPTER 11

INTRODUCTION

11.1. Probability theory and statistical experience. — In Chapter 2, probability theory has been characterized as a mathematical model for the description and interpretation of phenomena showing that particular type of regularity which we have called *statistical regularity*. So far we have only been concerned with the development of the purely mathematical theory of the model. We now proceed to consider the possibility of applying the model to actually observed phenomena, and to discuss the appropriate methods for making such applications. This discussion will form the subject of the remaining part of this book. In the present chapter some general introductory remarks will be given.

Suppose that we are investigating a certain phenomenon, and that we have at our disposal a set of *statistical data*, say in the form of a list of the recorded results of a series of repetitions of some experiment or observation connected with the phenomenon under investigation. We now want to find out whether this phenomenon may be adequately described by means of some suggested theoretical model involving probabilities. How should we proceed in order to test the agreement between the theoretical model and the actual observations as expressed by our data?

Consider the hypothesis that the suggested model is, in fact, adequate. On this hypothesis, we may compute the probabilities of various events connected with the experiment in question. Thus if A denotes some simple event which may occur, or fail to occur, at any performance of the experiment, we shall in general be able to find the numerical value P of the probability, according to our model, that A will occur. This theoretical probability has, in the field of statistical experience, its empirical counterpart in the frequency ratio f/n of A that has been observed in a series of n repetitions of the experiment. According to 2.3, the probability P should be regarded as a mathematical idealization of the corresponding frequency ratio f/n. *In order that our hypothesis may be accepted as satisfactory, we must evidently require that this correspondence between frequency ratios and probabilities should be borne out by our actual observations so that, in a long series of observations, the frequency ratio f/n should be approximately equal to P.*

In many important applications, we shall be concerned with the particular case when the probability P is a *small* number. In order to be able to accept the hypothesis that the model is adequate, we must then require that the actually observed frequency ratio f/n should be small when n is large. This means that A should be an *unusual* event, which in the long run will occur only in a small percentage of all repetitions of the corresponding experiment. We have already pointed out in 3.1 (cf. also 4.3, ex. 2) that in such a case we should be able to regard it as *practically certain* that A will not occur if we make one single performance (or even a small number of performances) of the experiment.

Any event A of sufficiently small probability P may thus be made to yield a *test* of the suggested model. We make one single performance of the experiment, and observe whether A occurs or not. In the former case we have obtained a result that would be highly unlikely under the hypothesis that the model is adequate, and so it will be natural to *reject* this hypothesis, and to regard it as disproved by experience. This is, of course, by no means equivalent to *logical* disproof. Even in a case when the model is in fact adequate, there will be the probability P that A may occur at our test performance, and thus we shall have the probability P of making a mistake by rejecting a true hypothesis. However, if P is sufficiently small, we may be willing to take this risk. — If, on the other hand, A fails to occur at the test performance, this is precisely what we should expect under the hypothesis that the model is adequate, and consequently we should feel inclined to say that, as far as the test goes, the hypothesis can be *accepted*. Again, this is not equivalent to a logical proof of the hypothesis. It only shows that, from the point of view of the particular test applied, the agreement of the suggested model with our observations is satisfactory. Before a probabilistic model can be regarded as definitely established. it will generally have to pass repeated tests of different kinds.

At a later stage (cf. in particular Chs. 14–15) we shall encounter various applications of this type of argument to important statistical questions. In the present section we shall only give one simple example.

Ex. — Suppose that, by means of the results (heads or tails) of repeated throws with a given coin, we want to decide whether the simple probabilistic model of independent observations (cf. 4.1) with a constant probability $P = \frac{1}{2}$ is applicable. One test of this model may be obtained in the following way.

Let us consider the performance of a series of 10,000 throws as one single experiment, and let A be the event that in this experiment the absolute frequency f of heads falls outside the interval $4{,}850 < f \le 5{,}150$. We have seen in 6.4 that, if our model is in fact adequate, the probability of A is approximately $P = 0.0027$. If we are willing to consider this probability so small that we may neglect the possibility of the occurrence of A in one single performance of the experiment, the argument given above will provide the following test: Perform a series of 10,000 throws, and observe the frequency f of heads.

If f falls outside the interval $4,850 < f \leq 5,150$, reject the hypothesis that the model is adequate, and otherwise accept (from the point of view of this particular test, as explained above).

If we adopt this rule, we shall have the probability 0.0027 of rejecting the hypothesis in a case when it is in fact true. Thus if we test in this way a large number of coins for each of which the model is in fact adequate, we shall be led to a wrong decision in about 2.7% of all cases. Obviously this percentage of wrong decisions will be reduced if the limits 4,850 and 5,150 of the interval of acceptance are widened, and vice versa.

11.2. Applications of probability theory. Description. — Suppose now that, after repeated tests of the kind discussed in the preceding section, we are ready to accept the suggested probabilistic model as giving a satisfactory mathematical theory of the phenomenon under investigation. Just as in the case of any other mathematical theory of some group of observed phenomena, we can then use our theory for making applications of various kinds. The applications of probability theory may be roughly classified under the headings: *Description, Analysis,* and *Prediction.* We shall make some brief general comments on each of these groups.

In the first place, the theory may be used for purely *descriptive* purposes. When a large set of statistical data has been collected, we want to describe, as a rule, the characteristic features of this material as briefly and concisely as possible. A first step will be to register the data in some convenient *table,* and perhaps to work out some *graphical representation* of the material. Moreover, we often want to be able to compute a small number of numerical quantities, which may serve as *descriptive characteristics* of our material, and which give in a condensed form as much as possible of the relevant information contained in the data.

Now in many cases the statistical data can be regarded as observed values of certain random variables, about the probability distributions of which we possess some information. The mathematical expressions of these probability distributions will then give us useful guidance with respect to the convenient arrangement of tabular and graphical representations. It will also be possible to calculate from the data estimated values of the parameters (mean values, central moments, correlation coefficients, etc.) appearing in the expressions of the probability distributions, and these estimated values will often render good service as descriptive characteristics.

The elements of tabulation and graphical representation of a statistical material, and of the numerical computation of some simple characteristics, will be given in Ch. 12. Further, some important properties of the characteristics will be discussed in Chs. 13–14.

In most cases, however, the final object of a statistical investigation will

157

not be of purely descriptive nature. The descriptive characteristics will, in fact, usually be required for some definite purpose. We may, e.g., want to compare different sets of data with the aid of the characteristics of each set, or we may want to form estimates of the characteristics that we might expect to find in other similar sets of data. An important instance of the latter situation comes up whenever our data are concerned with a sample of individuals from some population, and we want to use the observed characteristics of the sample to form estimates of the unknown characteristics of the total population. In all such cases, the description of the actual data forms only a preliminary stage of the inquiry, and we are in reality concerned with an application belonging to one of the two following classes.

11.3. Analysis. — The concepts and methods of probability theory provide powerful tools for a scientific *analysis* of statistical data. In the majority of investigations based on statistical material we are able to regard our data as observed values of random variables, and it is required to use the data for drawing certain inferences concerning the probability distributions of these variables.

The problems arising in this connection may belong to many different types. We shall indicate some of them by means of a few simple examples.

As our first example we take the general problem discussed in 11.1, where it was asked whether a given set of data could be adequately described by means of some suggested probabilistic model. In a particular case we may want to know, e.g., if a given set of measurements obtained in connection with some physical or biological experiment can be reasonably regarded as observed values of a normally distributed variable. A question of this type may be approached by working out some appropriate measure of the degree of agreement, or the *goodness of fit*, between the suggested distribution and our data, and then developing a test on the lines indicated in 11.1. This subject will be discussed in Ch. 15.

On the other hand, in many cases we are primarily interested in certain parameters connected with the distributions of our variables, such as the mean and the variance, the median and the quartiles, etc., and we want to *estimate* these unknown parameters by means of the observed data. In general we then possess some a priori information about the type of the distribution, e.g. that it is normal, or at least that it belongs to the continuous type, etc. A simple example arises when, assuming that a given machine has a fixed, though unknown, probability of producing a defective unit, we want to estimate the value of this probability from a series of trials. — Some general methods for obtaining estimates, and for assessing the value of various possible

158

estimates, will be studied in Ch. 14, while some particular problems belonging to this group will be treated in Ch. 16.

A highly important case of the estimation problem arises when, as already indicated in the preceding section, our data are concerned with a sample of individuals from some population, and we want to use the sample for drawing inferences about the total population. Say that we have drawn a sample from a batch of similar units produced in a mechanical factory, and that from the observed percentage of defective units in the sample we want to estimate the corresponding unknown percentage in the whole batch, or that from the observed incomes in a sample from some human population we want to estimate the unknown median income in the total population, etc. Some problems belonging to this group will be discussed in Ch. 16.

In problems of the types mentioned above it is often convenient to express the situation by saying, as in 11.1, that we want to use our data for *testing a hypothesis* concerning the probability distribution involved. Thus we may consider the hypothesis that a certain variable is normally distributed, that a certain probability is equal to $\frac{1}{2}$, that the sampled batch A has a percentage of defectives not greater than that of another batch B, etc. For any hypothesis of this kind, we want to find a criterion by means of which a given set of data will allow us to *accept* or *reject* the hypothesis (or possibly to decide against giving any definite verdict, and prefer to remain in doubt until further experience has been gained). This is a particular case of the general problem of *statistical decisions*, which arises whenever we are faced with a set of alternative decisions, one of which must be made on the basis of the information provided by a set of statistical data, and such that the degree of preference for the various possible decisions depends on the unknown probability distribution corresponding to the data. Some general comments on the problems of testing hypotheses and taking statistical decisions will be given in 14.6.

11.4. Prediction. — The word *prediction* should here be understood in a very wide sense, as related to the ability to answer questions such as: What is going to happen under given conditions? — What consequences are we likely to encounter if we take this or that possible course of action? — What course of action should we take in order to produce some desired event? — Prediction, in this wide sense of the word, is the ultimate *practical* aim of science.

Questions of these types will occur in many important practical problems, such as various problems of social and industrial planning, which will depend for their solution on the possibility of making *predictions of future statistical data*. Thus a rational planning of a school building or a telephone exchange will require some prediction of the future numbers of school children or telephone conversations, the planning of the financial system (premiums, funds, bonus rates and the like) of a life insurance company will depend on a prediction of future mortality rates, etc. If, from previous statistical experience, we possess some information about the probability distributions of the random variables entering into a question of this type, it will often be possible to reach at least a partial solution of the problem of statistical prediction.

Finally, it should be emphasized that there are no very sharp distinctions between the three main groups of applications here considered. The descriptive treatment of a statistical material will often be inseparably bound up with some estimation or comparison of analytical nature, and the taking of a statistical decision is often equivalent to a prediction. The whole classification is introduced here only as a convenient mode of presenting some important aspects of the applications of mathematical probability under various circumstances.

CHAPTER 12

DESCRIPTIVE TREATMENT OF A STATISTICAL MATERIAL

12.1. Tabulation and grouping. — In this chapter we shall assume throughout that we are concerned with a set of statistical data obtained by recording the results of a series of n repetitions of some given random experiment or observation. The repetitions are assumed to be mutually independent, and performed under uniform conditions.

As pointed out in 5.1, the result of each observation may be expressed in numerical form, i.e., by stating the observed values of a certain number of random variables. We shall first consider the case when there is only one variable X, so that our *statistical material* consists of n observed values, say $x_1, x_2,\ldots,x_n$, of the variable X.

The observed values are usually in the first instance registered on a list, or on written or punched cards. If the number n of observations does not exceed, say, 20 or 30, we can obtain a rough idea of the distribution simply by arranging the individual observed values in a table in order of increasing magnitude. Various graphical representations can then be drawn up, and numerical characteristics computed, directly from this table.

For larger materials, however, it soon becomes cumbersome to work directly with the individual observed values, and some kind of *grouping* is then usually performed as a preliminary step, before the further treatment of the data begins. The rules for grouping are different according as the random variable involved belongs to the discrete or the continuous type (cf. 5.2–5.3). There are, of course, also cases when variables of more general types appear in the applications, but these are relatively scarce and will not be considered here.

For a variable of the *discrete* type it is usually convenient to draw a table giving in the first column all values of the variable X represented in the material, and in the second column the *frequency f* with which each value of X has appeared during the n observations. A table of this kind will obviously give, in a conveniently abridged form, complete information about the distribution of the observed values. An example of such a table is shown in the two first columns of Table 12.1.1, which give the results of a count of the number of red blood corpuscles in each of the 169 compartments of a haemacytometer (data of N. G. HOLMBERG). Each compartment represents one observation, and the number of corpuscles in the compartment is the corresponding observed value of the variable X. As shown by the table, the smallest observed value of X was 4, which appeared in one single compartment, while the value 5 was observed in three compartments, and so on until the largest observed value 21, which appeared in one single compartment.

For a variable of the *continuous* type, the grouping procedure is slightly more complicated. Choose a suitable interval on the axis of X which contains all n observed values, and divide this interval into a certain number of *class intervals*. All observations belonging to the same class interval are then grouped together and counted, their number being denoted as the *class frequency* corresponding to the class interval. Then a table is drawn, where in the first column the limits of each class interval are specified, while the second column gives the corresponding class frequencies.

Two examples of such tables are shown in columns 3–6 of Table 12.1.1. Columns 3–4 give the results of a series of determinations of the percentage of ash in samples of coal, drawn from 250 different wagons (data of E. S. GRUMELL and A. C. DUNNINGHAM, British Standards Inst., Report No. 403), while columns 5–6 show the distribution according to income of 4,103 Swedish industrial leaders (Swedish census data, 1930). These examples call for comments in several respects.

Consider first the limits of the class intervals. In our table these have been given in the form that will usually be met with in practice, and in order to make it possible to ascertain the exact values of the limits, such a table should be accompanied by some information about the accuracy with which the original observations were made. Thus in the example with the ash percentages, it should be stated that the original data were given to the nearest unit in the second decimal place. The data were then grouped as

161

TABLE 12.1.1.

169 compartments, distributed according to number of blood corpuscles		250 samples of coal, distributed according to ash percentage		4,103 industrial leaders, distributed according to income	
Number of corpuscles	Frequency f	Percent ash	Frequency f	Income, unit 1000 kr.	Frequency f
4	1	9.00—9.99	1	0—1	49
5	3	10.00—10.99	3	1—2	86
6	5	11.00—11.99	3	2—3	144
7	8	12.00—12.99	9	3—4	189
8	13	13.00—13.99	13	4—6	477
9	14	14.00—14.99	27	6—8	401
10	15	15.00—15.99	28	8—10	350
11	15	16.00—16.99	39	10—20	1119
12	21	17.00—17.99	42	20—50	905
13	18	18.00—18.99	34	50—100	264
14	17	19.00—19.99	19	100—	119
15	16	20.00—20.99	14		
16	9	21.00—21.99	10		
17	6	22.00—22.99	4		
18	3	23.00—23.99	3		
19	2	24.00—24.99	—		
20	2	25.00—25.99	1		
21	1				
Total	169		250		4103

shown in the table, so that, e.g., the class interval 14.00–14.99 contains all observations registered by numbers from 14.00 to 14.99, both inclusive. However, when an observation is registered, e.g., as 14.27, this means that the really observed value lies somewhere between 14.265 and 14.275, and it follows that the exact limits of this class interval will be 13.995 and 14.995, the corresponding *class midpoint* being 14.445, and similarly for the other classes. — If the data hade been given to one place of decimals only, we should have had, e.g., the class 14.0–14.9, with the exact limits 13.95 and 14.95, and the class midpoint 14.45. — Sometimes the grouping of observations falling on the boundary between two class intervals is performed in a different way, every observation of, e.g., the value 14.00 being counted with half a unit in each of the two adjacent classes. The class intervals can then be taken as 13.00–14.00, 14.00–15.00, etc., the corresponding class midpoints being 13.50, 14.50, etc.

These observations will show that it is important to give, in connection with the tabulation of a grouped material, sufficient information about the method of grouping and the accuracy of the original observations.

Similar remarks apply to the income distribution example in Table 12.1.1, where the original data were given to the nearest integral number of kronor, and grouped as shown in the table. Find the exact limits of, say, the class 4000–5999!

It will be seen from Table 12.1.1 that the class intervals used for the ash percentage distribution are all of the same length. This is a useful practice, which gives a clear impression of the distribution of the material and simplifies various computations (cf. 12.5). As long as there are no definite reasons to the contrary, this rule should always be adhered to. However, in respects of strongly asymmetrical distributions, such as in the income distribution in the same table, it will often be necessary to use class intervals of varying length, as the table would become unwieldy if we should use throughout the whole range a class interval of the length that seems appropriate for the domain where the bulk of the observations lie. — The uppermost class interval in the income distribution is given in the table as 100,000 —, which means that it contains all observed cases with an income of at least 100,000 kronor. Such "open" classes appear sometimes at the beginning and the end of tables published in official statistics. They should be avoided whenever possible, and definite upper and lower limits given for each class.

By grouping a material in the way shown here, e.g., for the ash percentages we always lose some part of the information contained in the original data, since the table does not tell us the exact values of the original observations, but only the number of them belonging to each class. When grouped data are used for computations, it is customary to assume that all observations belonging to a given class are situated in the class midpoint. This approximation introduces an error, which obviously can be made as small as we please by choosing all class intervals sufficiently small, and thus reducing the loss of information due to the grouping. This, however, means a greater number of classes and an increasing size of the table, and so we shall give up something of the simplicity that was the main reason for grouping. When fixing our class intervals in practice, we shall thus have to take some middle course between the two opposite claims for accuracy and simplicity. As a simple practical rule it will be found that in many cases it is convenient to have a total number of classes between 10 and 20. With less than 10 classes we shall not get a very accurate description of the distribution, while the trouble to have more than 20 classes will hardly be worth while except for very large materials.

163

For the tabulation of a grouped material of simultaneous observations on two or more random variables we shall require tables with one argument for each variable. The rules for grouping are, mutatis mutandis, the same as in the case of one single variable. As an example we show in Table 12.1.2 the results of a series of 96 shots with an 8-mm machine gun, distributed according to azimuth and height deviations from the centre of the target (data from ARLEY and BUCH, Introduction to the probability and statistics, New York 1950, p. 168). The observations were made to the nearest mm and grouped in each variable so as to have the class midpoints given in the table. For each two-dimensional cell formed by the combination of both groupings, the table gives the corresponding *cell frequency*. The marginal totals of the table will, of course, show the distributions of the observations on each variable taken singly.

TABLE 12.1.2.

96 shots, distributed according to deviations from centre of target. (Class midpoints.)

Height deviation y	Azimuth deviation, x							Total
	-20	-10	0	10	20	30	40	
-50			1		2			3
-40		1	1	1	2			5
-30	1	1	3	5	2	1		13
-20	1	3	7	3	2	2		18
-10		2	6	10	3			21
0		1	6	6	6	1	1	21
10			3	3	3	1		10
20		1	1	2	1			5
Total	2	9	28	30	21	5	1	96

12.2 The sample values and their distribution. — Consider now a series of n independent repetitions of some random experiment, leading to the observed values $x_1,\ldots,x_n$ of the random variable X associated with the experiment.

In an important case already referred to in 11.2 and 11.3, the fundamental random experiment consists in drawing at random an individual from some population consisting of a finite number of individuals, and observing the value of some characteristic X that may vary from one individual to another. If the successive drawings are made with replacement (cf. 4.2), the observed values $x_1,\ldots,x_n$ will be independent, and will be denoted as a set of independent *sample values* of the variable X. (For the case of drawings without replacement, cf. 4.2 and 16.5.)

Even in the general case of independent observations it is often convenient to use a terminology borrowed from this special type of applications. Our n observations are then interpreted as a *random sample* from the *infinite population* of all observations that might have been made under the given circumstances, and the observed values $x_1,\ldots,x_n$ are looked upon as a set of independent *sample values* of the random variable X.

The variable X has a certain probability distribution, which may be known or unknown according to circumstances. This is the *distribution of the variable*, or the *distribution of the population*, or the *theoretical distribution* of the case. Like any probability distribution it can be defined by the corresponding d.f. $F(x)$, or (in the continuous case) by the fr.f. $f(x)$, and it can be visualized as a distribution of a unit of mass over the axis of X (cf. 5.1–5.3).

If the observed points $x_1, \ldots, x_n$ are marked on the axis of X, and the mass $1/n$ is placed in each point, we obtain another mass distribution, which may serve to represent the distribution of the sample values. This distribution will be denoted as the *distribution of the material*, or the *distribution of the sample*, or the *empirical distribution*. This will always be a distribution of the discrete type, having n discrete mass points (some of which may, of course, coincide). The mean, the variance and other characteristics of this distribution will be denoted as the *characteristics of the sample*, or the *observed characteristics*, as distinct from the corresponding characteristics *of the variable*, or *of the population*, which are associated with the distribution of the variable. We have thus defined the mean, the median, the variance, the skewness etc. of the sample, just as in 5.7 we have defined the corresponding characteristics of the variable.

Ex. — Suppose that the length of a screw made by a certain machine can be regarded as a normally distributed variable, with the mean $m = 30$ mm and the s.d. $\sigma = 0.25$ mm. Consider the random experiment which consists in drawing at random a screw produced by the machine, and measuring its length. According to the terminology introduced above, this is interpreted as the drawing of one individual from the infinite population of all screws that might have been made by the machine. Suppose that the experiment is repeated $n = 3$ times, and that in this way a sample of three screws measuring 29.79, 30.13, and 30.14 mm is obtained. The distribution of the *variable* is in this case a normal distribution with the mean 30 mm and the s.d. 0.25 mm, while the distribution of the *sample* is a discrete distribution having the mass $\frac{1}{3}$ in each of the points 29.79, 30.13, and 30.14. The mean and the s.d. of this distribution are

Mean: $\qquad \frac{1}{3}(29.79+30.13+30.14)=30.02$ mm,

S.D.: $\qquad \sqrt{\frac{1}{3}[(29.79-30.02)^2+(30.13-30.02)^2+(30.14-30.02)^2]}=0.16$ mm.

On the other hand, consider the three screws obtained above as a population of three individuals, from which we now draw (with or without replacement, as the case may be) a sample of two, and let us suppose that this sample consists of the two screws measuring 29.79 and 30.13 mm. In this case the distribution of the *population* is identical with the three-point distribution obtained in the preceding case, while the distribution of the *sample* is a two-point distribution with equal masses in the points 29.79 and 30.13.

In the case of simultaneous observations on two or more random variables, the distribution of the sample is defined in an analogous way. Thus if we have, e.g., a set of n observed pairs of values $(x_1, y_1), \ldots, (x_n, y_n)$ of two random variables X and Y, we mark the n corresponding points in a plane with the rectangular coordinates X and Y, and place the mass $1/n$ in each point. The resulting mass distribution is then, by definition, the distribution of the two-dimensional sample. The characteristics of this distribution, such as moments, coefficients of regression and correlation, etc., are the characteristics of the sample, or the observed characteristics. — If we are concerned with observations on three variables, the distribution of the sample will be a three-dimensional distribution, and so on.

12.3. Graphical representation. — Let $x_1, \ldots, x_n$ be n sample values of the random variable X, and let $F(x)$ denote the d.f. of this variable, which may be known or unknown. The distribution of the sample, as defined in the preceding section, has a certain d.f. that will be denoted by $F^*(x)$. Since the distribution of the sample is always of the discrete type, $F^*(x)$ will be a step function (cf. 5.2), with a step of the height $1/n$ in each of the observed points. If two sample values coincide, the height of the corresponding step will be $2/n$, etc. The function $F^*(x)$ is graphically represented by a broken line, which is known as the *sum polygon* of the sample.

Let x be a given number, and suppose that there are exactly f among our sample values that are $\leq x$. In the distribution of the sample, each of these points carries the mass $1/n$, so that the total mass situated in the domain $X \leq x$ is f/n, and thus we have (cf. 5.1)

$$F^* (x) = \frac{f}{n} .$$

Now f/n is the frequency ratio, in our series of n observations, of the event that consist in observing a value $\leq x$. The probability of this event can be expressed in terms of the d.f. of the variable:

$$P (X \leq x) = F (x).$$

If n is large, the frequency ratio f/n may be expected to be approximately

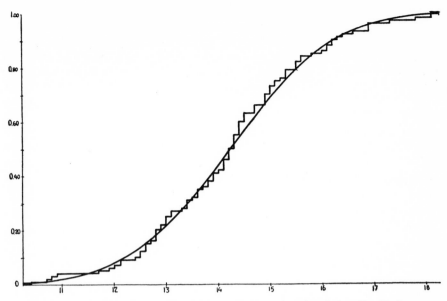

Fig. 24. Sum polygon for 100 mean temperatures (centigrades) in Stockholm, June 1841–1940, and normal distribution curve.

equal to the corresponding probability, so that we should have, *approximately*,

$$F^*(x) = F(x).$$

For large n, the sum polygon $y = F^*(x)$ can thus be expected to agree approximately with the distribution curve $y = F(x)$. Accordingly we may say, using the sampling terminology, that the distribution of the sample can be regarded as a *statistical image* of the distribution of the population.

As an example, we show in Fig. 24 the sum polygon of a sample of 100 observed values of the mean temperature in Stockholm for the month of June (the data are given in 15.3, ex. 3), together with a normal d.f. fitted to the observations (cf. 15.2), which may be taken as the hypothetical d.f. of the corresponding population.

For the graphical comparison between the sum polygon of a set of sample values and a normal d.f. it is, however, simpler to use a diagram drawn on graph paper with a transformed scale of the ordinates. Just as the ordinary log-paper uses the function $w = \log y$ for the scale of the ordinates, we can here use the inverse function $w = \Phi^{-1}(y)$, i.e., the unique root of the equation $y = \Phi(w)$, where $0 < y < 1$. *In a diagram with this scale, every normal d.f. becomes a straight line.* In fact, any normal distribution curve has an equation of the form $y = \Phi\left(\dfrac{x-m}{\sigma}\right)$, which gives $x = m + \sigma\Phi^{-1}(y) = m + \sigma w$. In the

167

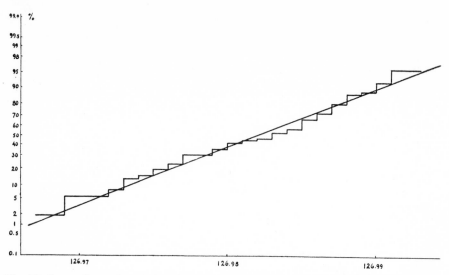

Fig. 25. Sum polygon for 52 determinations of the atomic weight of iodine, and normal distribution curve. The diagram is drawn on "probability paper", with the Φ^{-1} scale for the ordinates, so that the normal curve becomes a straight line.

(x, w) diagram, this equation represents a straight line. Fig. 25 shows the sum polygon of 52 experimental determinations of the atomic weight of iodine (cf. 15.2, ex. 3), together with the straight line corresponding to a normal d.f. fitted to the data. — Similar transformations may, of course, be used also in respect of other types of distributions.

If we have a grouped material, and the individual sample values are not available, the ordinates of the sum polygon are known only at the points where two class intervals meet. In such cases it is often better to use another kind of graphical representation. Let us take every class interval as the basis of a rectangle with the height $\dfrac{f}{nh}$, where h is the length of the interval, while f is the corresponding class frequency observed in our sample. The diagram obtained in this way is known as the *histogram* of the sample. The area of any rectangle in the histogram is equal to the corresponding frequency ratio f/n, and the total area of the rectangles corresponding to all the classes is thus equal to 1.

If the variable X is assumed to have a distribution of the continuous type, with the fr.f. $F'(x) = f(x)$, the probability that an observed value will fall within a specified class interval with the limits a and $a + h$ is

$$P(a < X < a + h) = \int_a^{a+h} f(x)\, dx.$$

168

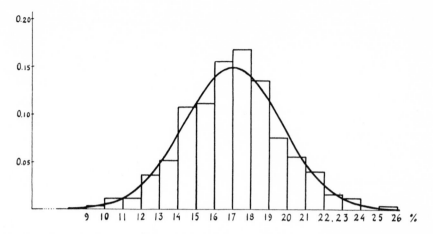

Fig. 26. Histogram for the ash percentage data of table 12.1.1, and normal frequency curve.

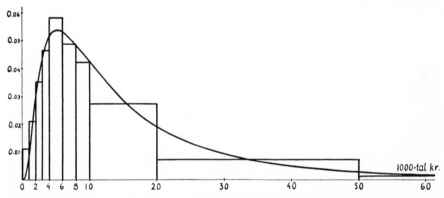

Fig. 27. Histogram for the income distribution of table 12.1.1, and log-normal frequency curve.

For large n, the frequency ratio f/n can be expected to be approximately equal to this probability, which means that the area of a given rectangle in the histogram should agree approximately with the area under the frequency curve $y = f(x)$ given by the second member in the last relation. Thus the upper contour of the histogram will form a statistical image of the frequency curve of the population, in the same sense as the sum polygon does so for the distribution curve. In Figs. 26 and 27, we show histograms for the ash percentage and income distributions from Table 12.1.1, compared with certain hypothetical frequency curves fitted to the data (cf. 15.2).

The factor n in the expression $\dfrac{f}{nh}$ for the height of a rectangle in the histogram can always be discarded, since this only amounts to a change of

169

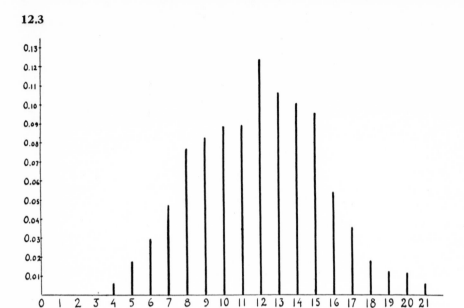

Fig. 28. Probability diagram for the blood corpuscle data of table 12.1.1.

scale on the vertical axis of the diagram. If all class intervals have the same length h, the whole denominator nh can be discarded, and the height of the rectangle will then simply be f. The total area of the histogram now becomes $h\Sigma f = nh$, and the comparison should be made with the curve $y = nh\,f(x)$ instead of $y = f(x)$. (What will be the scale on the vertical axis of Fig. 26 if the histogram is made in this way?).

When X is a variable of the discrete type, it is more convenient to use a diagram which, for every value of X represented in the material, has an isolated ordinate equal to the corresponding observed frequency ratio. This is the direct analogue of the probability diagram (cf. 5.2) of the discrete distribution of X, and for large n the corresponding ordinates should become approximately equal. Fig. 28 shows this *frequency diagram* for the blood corpuscle data from Table 12.1.1. This should be compared with the probability diagram for the POISSON distribution given in Fig. 15, p. 103.

Consider now a set of observed pairs of values $(x_1, y_1), \ldots, (x_n, y_n)$ of the random variables X and Y. When the individual observations are available, and n is not too large, the simplest graphical representation of this material is provided by the *dot diagram*, which is obtained by plotting each pair of observed values as a point in the (X, Y)-plane.

A grouped material with two variables can be graphically represented by a three-dimensional analogue of the histogram. If $a < X < a + h, b < Y < b + k$ is one of the rectangular cells used for the grouping, we take this as the basis

170

of a prism with the height $\dfrac{f}{nhk}$, where f is the corresponding cell frequency. The volume of the prism is then equal to the frequency ratio f/n, and the total volume of the prisms standing on all the cells is 1.

12.4. The characteristics of the sample. — In Chs. 5, 9, and 10 we have introduced standard notations for various classes of characteristics of random variables. In many of these cases Greek letters have been used, such as μ for central moments, σ for standard deviations, β and ϱ for coefficients of regression and correlation, etc. For the mean value of a variable, on the other hand, we have as a rule used the letter m.

In the sequel we shall often have occasion to consider certain characteristics of a sample, and at the same time also the corresponding characteristics of the variable (or the population). With respect to the notations, we shall in such cases as far as possible adhere to the following rules:

1. *The arithmetic mean of a number of observed quantities, such as $x_1, \ldots, x_n$, or $y_1, \ldots, y_k$, will be denoted by the corresponding letter with a bar:*

$$\bar{x} = \frac{1}{n}(x_1 + \cdots + x_n), \quad \bar{y} = \frac{1}{k}(y_1 + \cdots + y_k).$$

2. *When a certain population characteristic is ordinarily denoted by a Greek letter, the corresponding sample characteristic will be denoted by the corresponding Roman letter, in italic type, thus m for μ, s for σ, b and r for β and ϱ, etc.*

If our data consist of n observed values $x_1, \ldots, x_n$ of a random variable X, the distribution of the material (cf. 12.2) is a discrete distribution with the mass $1/n$ in each point x_i. The *mean* of the sample is then by 5.5.1

$$\bar{x} = \frac{1}{n}\sum_i x_i,$$

where the summation is extended over all sample values, $i = 1, 2, \ldots, n$. The *variance*, or the square of the *standard deviation*, of the sample is by 5.7.1

$$s^2 = \frac{1}{n}\sum_i (x_i - \bar{x})^2$$

$$= \frac{1}{n}\sum_i x_i^2 - \bar{x}^2.$$

171

More generally, the *moment* and the *central moment* of order ν of our sample are

$$a_\nu = \frac{1}{n} \sum_i x_i^\nu, \quad m_\nu = \frac{1}{n} \sum_i (x_i - \bar{x})^\nu.$$

The relations 5.6.3 and 5.6.4 between the moments and the central moments hold for any distribution, and hence in particular for the distribution of our sample. It follows that if in these relations m is replaced by $\bar{x}$, α_ν by a_ν, and μ_ν by m_ν, we obtain valid relations between the moments and the central moments of the sample (cf. 12.5.1).

The measures of *skewness* and *excess* of the sample are in accordance with 5.7.8 and 5.7.9

$$g_1 = \frac{m_3}{s^3}, \quad g_2 = \frac{m_4}{s^4} - 3.$$

The *median* z_2 of the sample is, by definition (cf. 5.7), the value that occupies the central position when the sample values are arranged in order of increasing magnitude. When n is odd, the median is uniquely determined. When n is even, on the other hand, every point in the interval between the two central sample values satisfies the median condition (cf. 5.7), and usually the midpoint of this interval is taken as the sample median. — The first and third sample *quartiles* z_1 and z_3 are defined in a similar way.

All these quantities are characteristics of the distribution of the sample, and may serve as measures of location, dispersion, skewness, etc., for this distribution, just as the corresponding population characteristics may be used in this way for the distribution of the population. Thus, e.g., the mean $\bar{x}$ usually provides a good measure of location for the sample. When the distribution of the sample is strongly asymmetric, however, the extreme values situated far out in the tails will have a dominating influence on the value of the mean. As already indicated in 5.7, the median may in such cases be preferable as location measure. Thus for the income distribution in Table 12.1.1 we find the mean $\bar{x} = 23{,}448$ kr. while the median is $z_2 = 13{,}177$ kr.[1] A look at the histogram in Fig. 27 will show that in this case the median gives a more satisfactory location than the mean, which is strongly influenced by the small number of cases with very large incomes.

The *greatest* and the *smallest* of the sample values will be denoted by x_{max} and x_{min}. The difference $x_{max} - x_{min}$ is called the *range* of the sample, and may often serve as a useful dispersion measure.

---

[1] These are exact values, computed from the ungrouped data not given in our table.

For a sample of n observed pairs of values $(x_1, y_1), \ldots, (x_n, y_n)$ of the two random variables X and Y we have the two mean values

$$\bar{x} = \frac{1}{n} \sum_i x_i, \quad \bar{y} = \frac{1}{n} \sum_i y_i,$$

and the second order central moments (cf. 9.1.1)

$$m_{20} = s_1^2 = \frac{1}{n} \sum_i (x_i - \bar{x})^2 = \frac{1}{n} \sum_i x_i^2 - \bar{x}^2,$$

$$m_{11} = \frac{1}{n} \sum_i (x_i - \bar{x})(y_i - \bar{y}) = \frac{1}{n} \sum_i x_i y_i - \bar{x}\,\bar{y},$$

$$m_{02} = s_2^2 = \frac{1}{n} \sum_i (y_i - \bar{y})^2 = \frac{1}{n} \sum_i y_i^2 - \bar{y}^2.$$

The *correlation coefficient* of the sample is by 9.1.4

$$r = \frac{m_{11}}{\sqrt{m_{20}\, m_{02}}} = \frac{m_{11}}{s_1\, s_2},$$

while the two regression coefficients (cf. 9.2.6) are

$$b_{21} = \frac{r\, s_2}{s_1}, \quad b_{12} = \frac{r\, s_1}{s_2}.$$

For samples of more than two simultaneously observed variables, regression coefficients, partial and multiple correlation coefficients, etc., are defined in a similar way. All relations between these various coefficients given in Ch. 10 will hold good if the population characteristics are throughout replaced by the corresponding sample characteristics.

12.5. Numerical computation of characteristics.

— If the individual data are available, and the number of observations is not too large, all sample characteristics mentioned in the preceding section can be directly computed by means of the definitions and formulae there given. We shall now consider some methods of computation that may be used in cases where these conditions are not satisfied. *Here, as everywhere in the sequel, the reader is expected to work through all numerical examples in detail.*

Consider first the case when we have a sample of values of a discrete variable, given in the form of a frequency table such as the one shown for the blood corpuscle observations in Table 12.1.1. If the values of the variable represented in the table are denoted by $\xi_1, \ldots, \xi_k$, and the corresponding frequencies are $f_1, \ldots, f_k$, we obtain the following expressions for the mean, the moments and the central moments of the sample

$$\bar{x} = \frac{1}{n}\sum f_i \xi_i, \quad a_\nu = \frac{1}{n}\sum f_i \xi_i^\nu, \quad m_\nu = \frac{1}{n}\sum f_i (\xi_i - \bar{x})^\nu,$$

where the summation is extended over the values $i = 1, 2, \ldots, k$. For the numerical computation of these quantities we start from the frequency table with the two columns giving the ξ_i and the f_i. By repeated multiplication of each f_i with the corresponding ξ_i we obtain further columns with the products $f_i \xi_i$ and $f_i \xi_i^2$. These are sufficient for the computation of the mean and the variance. If the central moments m_3 and m_4 are also required, the procedure has to be continued so as to obtain also the products $f_i \xi_i^3$ and $f_i \xi_i^4$. (Higher central moments can be found in the same way, but they will practically never be required.) Dividing the column totals by n, we then obtain the mean $\bar{x}$ and the moments a_ν, according to the above expressions. The central moments are then found from the relations (cf. 5.6.3 and 5.6.4)

$$m_2 = s^2 = a_2 - \bar{x}^2,$$

(12.5.1)
$$m_3 = a_3 - 3\bar{x} a_2 + 2\bar{x}^3,$$

$$m_4 = a_4 - 4\bar{x} a_3 + 6\bar{x}^2 a_2 - 3\bar{x}^4.$$

One-dimensional characteristics such as $\bar{x}$ and s should as a rule be given to one or at most two places of decimals more than the original data. Dimensionless constants, such as measures of skewness, coefficients of correlation and regression, etc., should be given with two or at most three significant figures. In order to reach this accuracy in the final values it will often be necessary to have at least two or three figures more in the values of $\bar{x}$, a_2 etc. used in 12.5.1.

A valuable check on the computation is obtained by adding a column containing the products $f_i(\xi_i + 1)^\nu$, where ν is the order of the highest moment required, and then checking the column totals by means of the relation

$$\sum f_i (\xi_i + 1)^\nu = \sum f_i + \binom{\nu}{1}\sum f_i \xi_i + \cdots + \sum f_i \xi_i^\nu.$$

As an example we show in Table 12.5.1 the computation of moments for the blood corpuscle observations in Table 12.1.1. We have here, as well as in the examples given below, included the columns required for the computation of the third and fourth order moments, though in practice in many cases only

174

TABLE 12.5.1.

ξ	f	$f\xi$	$f\xi^2$	$f\xi^3$	$f\xi^4$	$f(\xi+1)^4$
4	1	4	16	64	256	625
5	3	15	75	375	1,875	3,888
6	5	30	180	1,080	6,480	12,005
7	8	56	392	2,744	19,208	32,768
8	13	104	832	6,656	53,248	85,293
9	14	126	1,134	10,206	91,854	140,000
10	15	150	1,500	15,000	150,000	219,615
11	15	165	1,815	19,965	219,615	311,040
12	21	252	3,024	36,288	435,456	599,781
13	18	234	3,042	39,546	514,098	691,488
14	17	238	3,332	46,648	653,072	860,625
15	16	240	3,600	54,000	810,000	1,048,576
16	9	144	2,304	36,864	589,824	751,689
17	6	102	1,734	29,478	501,126	629,856
18	3	54	972	17,496	314,928	390,963
19	2	38	722	13,718	260,642	320,000
20	2	40	800	16,000	320,000	388,962
21	1	21	441	9,261	194,481	234,256
Total	169	2,013	25,915	355,389	5,136,163	6,721,430

the mean and the variance will be wanted, which of course shortens the work considerably. The reader should go through all the calculations in detail.

The column totals are first checked by the aid of the above relation. Dividing by $n = 169$ we then obtain $\bar{x} = 11.91124$, $a_2 = 153.3432$, $a_3 = 2,102.893$ and $a_4 = 30,391.50$. Computing each term in the last members of the relations 12.5.1 with seven significant figures, we then find $m_2 = 11.466$, $m_3 = 3.25$ and $m_4 = 347.3$. From these values we can obtain the standard deviation s and the measures of skewness and excess, g_1 and g_2, and finally the result of the whole computation may be summed up in the values

$$\bar{x} = 11.91, \quad s = 3.39, \quad g_1 = 0.084, \quad g_2 = -0.358.$$

For a material of this type, the median and the quartiles are found directly by inspection of the frequency table, and we obtain in the example here considered

$$z_1 = 9, \quad z_2 = 12, \quad z_3 = 14.$$

We now proceed to the case of a grouped material from a continuous distribution. Since in this case the individual data are unknown, the exact values of the moments etc. cannot be found. As indicated in 12.1, we shall then have recourse to the approximation which consists in assuming that every observed value is situated in the midpoint of the class interval to which it belongs. If the class midpoints are ξ_i, and the corresponding class frequencies f_i, the moments and central moments will have the same expressions as in the preceding case, though these are now only valid as approximations.

If all class intervals have the same length h, we can save a considerable amount of work by a simple transformation. Instead of numbering the classes from 1 to k, we arbitrarily choose one of the class intervals—preferably one of the central classes of the material—and give it the number *zero*. We then start the numbering from this origin, giving negative numbers to the class intervals on the left, and positive to those on the right. If ξ_0 is the midpoint of the class zero, the midpoint of the class carrying the number i will then be

$$\xi_i = \xi_0 + i\,h,$$

so that we have

$$\bar{x} = \frac{1}{n}\sum_i f_i\,(\xi_0 + i\,h),$$

$$m_\nu = \frac{1}{n}\sum_i f_i\,(\xi_0 + i\,h - \bar{x})^\nu.$$

Writing

$$a'_\nu = \frac{1}{n}\sum_i f_i\,i^\nu,$$

this gives us

$$\bar{x} = \xi_0 + a'_1\,h,$$

$$m_2 = s^2 = (a'_2 - a'^2_1)\,h^2,$$

$$m_3 = (a'_3 - 3\,a'_1\,a'_2 + 2\,a'^3_1)\,h^3,$$

$$m_4 = (a'_4 - 4\,a'_1\,a'_3 + 6\,a'^2_1\,a'_2 - 3\,a'^4_1)\,h^4.$$

As already mentioned, these expressions give only approximate values of the characteristics, since we have replaced all observed values by the corresponding class midpoints. In certain cases the error introduced in this way can be reduced by applying the formulae known as SHEPPARD's corrections. For the deduction of these formulae we refer to *Mathematical Methods*, p. 359. They should be applied in cases where the histogram indicates a contact of high order between the X axis and both tails of the frequency curve, as e.g. in the distribution shown in Fig. 26. When this condition is not satisfied, as e.g. in the case of Fig. 27, it is better to use uncorrected moments. For $\bar{x}$ and m_3 there are no correction terms, while for m_2 and m_4 we have

TABLE 12.5.2.

ξ	i	f	fi	fi^2	fi^3	fi^4	$f(i+1)^4$
9.495	-8	1	-8	64	-512	4096	2401
10.495	-7	3	-21	147	-1029	7203	3888
11.495	-6	3	-18	108	-648	3888	1875
12.495	-5	9	-45	225	-1125	5625	2304
13.495	-4	13	-52	208	-832	3328	1053
14.495	-3	27	-81	243	-729	2187	432
15.495	-2	28	-56	112	-224	448	28
16.495	-1	39	-39	39	-39	39	0
17.495	0	42	0	0	0	0	42
18.495	1	34	34	34	34	34	544
19.495	2	19	38	76	152	304	1539
20.495	3	14	42	126	378	1134	3584
21.495	4	10	40	160	640	2560	6250
22.495	5	4	20	100	500	2500	5184
23.495	6	3	18	108	648	3888	7203
24.495	7	0	0	0	0	0	0
25.495	8	1	8	64	512	4096	6561
Total		250	-120	1814	-2274	41330	42888

$$m_2 \text{ (corr)} = m_2 - \frac{1}{12} h^2,$$

$$m_4 \text{ (corr)} = m_4 - \frac{1}{2} m_2 h^2 + \frac{7}{240} h^4.$$

As an example we show in Table 12.5.2 the computations for the ash percentage distribution of Table 12.1.1, where we have chosen the class 17.00–17.99 as zero class, so that $\xi_0 = 17.495$, while the common length of the class intervals is $h = 1$. The last column gives a check of the column totals in the same way as in the preceding case.

After checking the column totals we divide by $n = 250$ and obtain $a_1' = -0.480$, $a_2' = 7.256$, $a_3' = -9.096$, $a_4' = 165.320$. According to the formulae given above this gives $\bar{x} = 17.015$, $m_2 = 7.0256$, $m_3 = 1.1315$, $m_4 = 157.7271$. Finally the SHEPPARD corrections lead to the corrected values m_2 (corr.) $= 6.9423$ and m_4 (corr.) $= 154.2435$. The result of the computations can then be summed up in the values

$$\bar{x} = 17.015, \ s = 2.635, \ g_1 = 0.062, \ g_2 = 0.200.$$

It will be seen that in this example the main part of the work was spent on the computation of the sums $\Sigma f_i\, i^v$. Often this task can be simplified by using an *additive* method of computation instead of the *multiplicative* method described above. In particular the additive method should be recommended when a writing calculating machine is available. The additive method can be modified in various ways, and we shall here only show how it may be applied as an alternative method of computing the sums $\Sigma f_i\, i^v$ in the case treated by the multiplicative method in Table 12.5.2.

The work can be arranged as shown in Table 12.5.3. Starting from the f column in the frequency table, we add the f_i from above until the row $i = -1$, and from below until the row $i = +1$, noting the value of the sum obtained at each step. In this way we obtain the column headed Σ_0, which is then treated in the same way to form the column Σ_1, and so on. If only the mean and the variance are required, we may stop after working out the column Σ_2, while Σ_3 and Σ_4 will be necessary for the computation of m_3 and m_4. Let us now denote by $A_1, \ldots, A_4$ the numbers occupying the row $i = +1$ in the columns $\Sigma_1, \ldots, \Sigma_4$, while $B_1, \ldots, B_4$ are the numbers on the row $i = -1$ in the same columns. By some calculation we then obtain the expressions

$$\Sigma f_i\, i = A_1 - B_1,$$

$$\Sigma f_i\, i^2 = 2\, A_2 + 2\, B_2 - A_1 - B_1,$$

$$\Sigma f_i\, i^3 = 6\, A_3 - 6\, B_3 - 6\, A_2 + 6\, B_2 + A_1 - B_1,$$

$$\Sigma f_i\, i^4 = 24\, A_4 + 24\, B_4 - 36\, A_3 - 36\, B_3 + 14\, A_2 + 14\, B_2 - A_1 - B_1.$$

In Table 12.5.3 the numbers A_j and B_j have been printed in italics. The reader should substitute these numbers in the above expressions, and verify that this procedure leads to the same values of the sums $\Sigma f_i\, i^v$ as the multiplicative method used above. The central moments etc. are then found in the same way as above.

When the additive method is used, it is often advantageous to assign the class number zero to the *lowest* class in the frequency table. In the above example we should then have $i = 0$ instead of -8, $i = 1$ instead of -7, etc. The sums B_j then disappear, and the expressions for the sums $\Sigma f_i\, i^v$ are considerably simplified.

The *median* and the *quartiles* of a grouped sample can only be found approximately by interpolation. For the ash percentage data discussed above the median lies between the 125:th and the the 126:th value when the observations are ranked in order of magnitude. Now it is seen that there are 123 observed values below the lower limit 16.995 of the class zero. Assuming that the 42 observations in the class zero are uniformly distributed over the

TABLE 12.5.3.

i	f	Σ_0	Σ_1	Σ_2	Σ_3	Σ_4
-8	1	1	1	1	1	1
-7	3	4	5	6	7	8
-6	3	7	12	18	25	33
-5	9	16	28	46	71	104
-4	13	29	57	103	174	278
-3	27	56	113	216	390	668
-2	28	84	197	413	803	1471
-1	39	123	320	733	1536	3007
			(B_1)	(B_2)	(B_3)	(B_4)
0	42					
			(A_1)	(A_2)	(A_3)	(A_4)
1	34	85	200	434	878	1677
2	19	51	115	234	444	799
3	14	32	64	119	210	355
4	10	18	32	55	91	145
5	4	8	14	23	36	54
6	3	4	6	9	13	18
7	0	1	2	3	4	5
8	1	1	1	1	1	1

class interval of length $h = 1$, we find that the median should lie between $16.995 + \dfrac{2}{43} = 17.042$ and $16.995 + \dfrac{3}{43} = 17.065$. Approximate values of the quartiles can be found in the same way.

Consider now the case of a grouped two-dimensional sample. such as the one given in Table 12.1.2. We first compute the mean values $\bar{x}$ and $\bar{y}$ and the standard deviations s_1 and s_2 for each variable separately. These values are found from the distributions of the X and Y values given by the marginal totals of the table. Then the product moment m_{11} of the sample is computed from the expression

$$m_{11} = \frac{hk}{n^2} \left(n \sum_{i,j} f_{ij}\, ij - \sum_i f_{i.}\, i \cdot \sum_j f_{.j}\, j \right),$$

where h and k are the lengths of the class intervals for X and Y respectively, i and j are the class numbers, f_{ij} is the observed frequency for the rectangular cell formed by the i-th and j-th class intervals, and finally $f_{i.} = \sum_j f_{ij}$ and $f_{.j} = \sum_i f_{ij}$ are the marginal totals. The sums $\sum_i f_{i.}\, i$ and $\sum_j f_{.j}\, j$ are the same as those used for the computation of the means $\bar{x}$ and $\bar{y}$,

179

while the double sum $\sum_{i,j} f_{ij} i j$ must be computed term by term and summed over all the cells in the two-dimensional table. — The correlation coefficient r and the regression coefficients b_{21} and b_{12} can now be found from the formulae of 12.4.

For the shot material of Table 12.1.2 we obtain in this way from the marginal totals

$$\bar{x} = 8.125, \quad \bar{y} = -11.563, \quad s_1 = 11.209, \quad s_2 = 16.414,$$

where s_1 and s_2 have been found after applying the SHEPPARD corrections to the "raw" variances m_{20} and m_{02}. Using the expression given above we then find, assigning on each axis the class number zero to the class containing the origin,

$$\sum_{i,j} f_{ij} i j = -75, \quad \sum_i f_{i.} i = 78, \quad \sum_j f_{.j} j = -111,$$

and accordingly $m_{11} = 15.820$. By means of these values we obtain

$$r = 0.086, \quad b_{21} = 0.126, \quad b_{12} = 0.059.$$

Finally we shall consider the following example of a three-dimensional sample. For each of $n = 287$ boys the length X_1 and the weight X_2 of the body were observed, as well as the best achievement in jumping, X_3. The following characteristics were obtained for each variable taken separately (data from E. ABRAMSON, Hygienisk Tidskr. 1945, p. 295):

	Sample mean	Sample s.d.
Length	162.6 cm	10.64 cm
Weight	49.34 kg	9.67 kg
Jump	437.3 cm	60.4 cm

The total correlation coefficients of the sample were

$$r_{12} = 0.876, \quad r_{13} = 0.571, \quad r_{23} = 0.705.$$

By means of these values we find from 10.2.1 the partial correlation coefficients

$$r_{12\cdot3} = 0.813, \quad r_{13\cdot2} = -0.136, \quad r_{23\cdot1} = 0.517,$$

and from 10.3.1 the multiple correlation coefficient between jumping ability on the one side, length and weight on the other:

$$r_{3(12)} = 0.712.$$

The sample regression plane of jumping ability on length and weight has according to 10.1.2 the equation

$$x_3 = 350.6 - 1.135\,x_1 + 5.497\,x_2.$$

The reader should work out all these computations in detail.

CHAPTER 13

SAMPLING DISTRIBUTIONS

13.1. The characteristics as random variables. — Let X be a random variable with the d.f. $F(x)$, and consider a series of n independent repetitions of the random experiment to which X is attached.

Before any experiment has been made, the values of X that will be observed may be conceived as n independent random variables $X_1, \ldots, X_n$, each having the same d.f. $F(x)$. Any function of these variables, say $g(X_1, \ldots, X_n)$, will then itself be a random variable, with a certain probability distribution determined by the functions F and g.

Performing the series of n repetitions, we obtain a sample of n observed values of X, which as usual we denote by $x_1, \ldots, x_n$. Any sample characteristic will be a function of the sample values, say $g(x_1, \ldots, x_n)$. Now this may be regarded as an observed value of the random variable $g(X_1, \ldots, X_n)$, and accordingly the probability distribution of this latter variable will be called the *sampling distribution* of the characteristic $g(x_1, \ldots, x_n)$.

If samples of n values are repeatedly drawn in this way, and if for each sample the characteristic $g(x_1, \ldots, x_n)$ is computed, we shall obtain a sequence of observed values of the random variable $g(X_1, \ldots, X_n)$. The sum polygon or the histogram of this sequence of values will then (cf. 12.3) provide a statistical image of the distribution curve or the frequency curve corresponding to the sampling distribution of the characteristic g.

In this way every sample characteristic is associated with a certain random variable, and we may talk of the sampling distribution of the mean $\bar{x} = \frac{1}{n}\sum x_i$, of the variance $s^2 = \frac{1}{n}\sum (x_i - \bar{x})^2$, of the median, etc. The sampling distribution of a given characteristic is in principle determined by the d.f. $F(x)$ of the basic random variable X, even though it may sometimes be difficult to find an explicit expression in terms of $F(x)$.

The properties of sampling distributions are of great importance for various applications, and some of the simplest cases will be studied in the present chapter.

181

13.2. The moments of the characteristics. — For the characteristics of the basic random variable X we shall use the notations introduced in Ch. 5. Thus m and σ will denote the mean and the s.d. of X, while μ_ν is the central moment of order ν, etc. We shall assume throughout that all moments appearing in our formulae are finite.

Consider first the simplest sample characteristic, the *mean*

$$\bar{x} = \frac{1}{n} \sum_i x_i.$$

If $x_1, \ldots, x_n$ are regarded as independent random variables[1], each with the same d. f. $F(x)$, the mean $\bar{x}$ will also be a random variable, and we obtain from 5.7.6 and 5.7.7

(13.2.1) $$E(\bar{x}) = m, \quad D(\bar{x}) = \frac{\sigma}{\sqrt{n}}.$$

Thus the mean value of the sample characteristic $\bar{x}$ is equal to the corresponding characteristic of the variable, or population characteristic, m. Moreover, the s.d. $\sigma/\sqrt{n}$ of $\bar{x}$ will be small for large values of n, and according to TCHEBYCHEFF's theorem (cf. 5.8) the mass in the sampling distribution of $\bar{x}$ will then be concentrated in the vicinity of the point m. Thus for sufficiently large values of n we may be practically certain that the sample mean $\bar{x}$ will be approximately equal to its mean value, which is the population mean m. If m is unknown, we can then use $\bar{x}$ as an *estimate* of m.

Let us now consider the *variance* of the sample

$$s^2 = m_2 = \frac{1}{n} \sum_i (x_i - \bar{x})^2 = \frac{1}{n} \sum_i (x_i - m)^2 - (\bar{x} - m)^2.$$

Since we have

$$E(x_i - m)^2 = \sigma^2, \quad E(\bar{x} - m)^2 = \frac{\sigma^2}{n},$$

it follows that

(13.2.2) $$E(s^2) = \sigma^2 - \frac{\sigma^2}{n} = \frac{n-1}{n} \sigma^2.$$

Thus the mean of the sample characteristic s^2 is *not*, as in the case of $\bar{x}$, equal to the corresponding population characteristic σ^2, but to $\frac{n-1}{n} \sigma^2$.

---

[1] For the sake of simplicity, we shall from now on often use the same letters, e.g. $x_1, \ldots, x_n$, to denote a set of observed values and the corresponding random variables. This will hardly give rise to any confusion.

It is true that the difference between σ^2 and $\dfrac{n-1}{n}\sigma^2$ is insignificant when n is large, but for moderate values of n it will be preferable to consider the *corrected* sample variance

$$\frac{n}{n-1}\,s^2 = \frac{1}{n-1}\sum_i (x_i - \bar{x})^2,$$

which has a mean value exactly equal to σ^2, as shown by the relation

$$E\left(\frac{n}{n-1}\,s^2\right) = \frac{n}{n-1}\cdot\frac{n-1}{n}\,\sigma^2 = \sigma^2.$$

The standard deviation of s^2 is given by the expression

$$(13.2.3) \qquad D(s^2) = \sqrt{\frac{\mu_4 - \mu_2^2}{n} - \frac{2(\mu_4 - 2\,\mu_2^2)}{n^2} + \frac{\mu_4 - 3\,\mu_2^2}{n^3}},$$

for the proof of which we refer to *Mathematical Methods*, p. 348. For large n, the s.d. will again be small, and s^2 can then be expected to agree approximately with the population variance σ^2 since, as already pointed out, the mean of s^2 is practically equal to σ^2 when n is large.

By similar, though often somewhat intricate, calculations we can find the mean value and the s.d. of any moment characteristic of the sample. The results are largely of the same type as in the cases of $\bar{x}$ and s^2. The mean value of a certain sample characteristic will be—exactly, as in the case of $\bar{x}$, or at least asymptotically for large n, as in the case of s^2—equal to the corresponding population characteristic, and the s.d. will be small for large values of n. It thus follows that, at least for large n, the sample characteristics may serve as estimates of the corresponding population characteristics.

13.3. Asymptotic sampling distribution. — It follows from the central limit theorem (cf. 7.4) that for large n the sample mean $\bar{x} = \dfrac{1}{n}\sum_i x_i$ is asymptotically normally distributed $(m,\ \sigma/\sqrt{n})$. As pointed out in 7.4, the same property will hold also for other symmetric functions of the sample values. In fact, it can be proved (cf. *Mathematical Methods*, Ch. 28) that under fairly general conditions any moment or central moment of the sample, as well as any algebraic function of such moments, is asymptotically normally distributed for large n. This includes, e.g., the sample s.d. s, the measures of skewness and excess, g_1 and g_2, etc., and also certain characteristics not related to moments, such as the median and the quartiles. Moreover, the theorem can be extended to samples from multi-dimensional distributions. Thus, e.g., for a two-dimensional sample the correlation coefficient r and the regression co-efficients b_{21} and b_{12} are asymptotically normal.

If z_i is one of the sample quartiles ($i = 1, 2, 3$), and ζ_i is the corresponding quartile of the population, which is assumed to belong to the continuous type, it can be shown (cf. *Mathematical Methods*, Chapter 28) that z_i is asymptotically normal

$$\left(\zeta_i, \frac{\sqrt{i(4-i)}}{4 f(\zeta_i) \sqrt{n}} \right),$$

where $f(x)$ denotes the frequency function of the distribution.

There are, however, some exceptions from the general rule. Thus the largest and smallest sample values and the sample range are as a rule not even approximately normally distributed. Further, the multiple correlation coefficient may under certain circumstances form an exception. Apart from these cases, all sample characteristics mentioned in this book are included in the general theorem of asymptotically normal distribution.

Evidently this property is of great importance for the applications. When a certain variable is known to be normally distributed, its distribution is completely characterized by the mean and the s.d. We know, e.g., that the probability that an observed value of such a variable will differ from its mean by more than three times the s.d. is only 0.27%, and according to the general remarks of 11.1 and 11.3 this will enable to us test the agreement between an observed sample characteristic and some hypothetic value of the corresponding mean. In the next chapter we shall meet with several applications of this type of argument.

How many observations should we require in a sample, in order that we may be justified in assuming the sample characteristics to be normally distributed? The answer to this question will largely depend on the type of characteristics concerned. When we are dealing with means of samples from distributions that are not extremely skew, the approximation will usually be sufficient for ordinary practical purposes as soon as $n > 30$. For variances, medians, coefficients of skewness and excess, correlation coefficients numerically smaller than 0.2, etc., it is advisable to require that n should be at least about 100. For correlation coefficients considerably different from zero, even samples of 300 do not always give a satisfactory approximation.

13.4. Exact sampling distributions. — In the two preceding sections we have studied the moments of some simple sample characteristics, and the asymptotic properties of sampling distributions for large samples. With respect to the d.f. $F(x)$ of the corresponding population, we have so far only assumed that all moments involved in our considerations are finite.

In order to reach more precise results about the properties of sampling distributions, it will be indispensable to introduce further assumptions about $F(x)$. Suppose, e.g., that we want to know whether it is possible to find reasonably simple *explicit expressions* for the sampling distributions of the principal characteristics. If we are to attack this question with some chance of success, it will be necessary to start from rather special assumptions about the mathematical form of $F(x)$. The case that has been most thoroughly studied is the case when $F(x)$ is a *normal d.f.*, and this will be considered in the present section.

Although special, the case of a normal $F(x)$ is of great importance for the applications. As we have pointed out in 7.7, many variables which express the results of industrial, biological or medical experiments are at least approximately normally distributed, and whenever we are concerned with variables of this type the results given below will be at least approximately true.

If the variable X is normal (m, σ), the sample mean

$$\bar{x} = \frac{1}{n} \sum_i x_i$$

will according to 7.3 be normal $(m, \sigma/\sqrt{n})$. The validity of the normal distribution is here exact for any given value of n, and not only asymptotical for large n, as in the case discussed in the preceding section, where no assumptions were made about $F(x)$.

We shall now try to find the distribution of the sample variance

$$s^2 = \frac{1}{n} \sum_i (x_i - \bar{x})^2.$$

If the variable X is replaced by $X - m$, and consequently every x_i is replaced by $x_i - m$, it will be seen that the value of s^2 is not affected. Thus we may assume $m = 0$ without restricting the generality of our considerations. The random variables $x_1, \ldots, x_n$ will then be independent and normal $(0, \sigma)$. We now introduce new variables $y_1, \ldots, y_n$ by the substitution

$$y_1 = \frac{1}{\sqrt{1 \cdot 2}} (x_1 - x_2),$$

$$y_2 = \frac{1}{\sqrt{2 \cdot 3}} (x_1 + x_2 - 2 x_3),$$

(13.4.1)   . . . .

$$y_{n-1} = \frac{1}{\sqrt{(n-1)\,n}}\,(x_1 + x_2 + \cdots + x_{n-1} - (n-1)\,x_n),$$

$$y_n = \frac{1}{\sqrt{n}}\,(x_1 + x_2 + \cdots + x_{n-1} + x_n).$$

We shall first show that the y_i are, like the x_i, independent and normal $(0, \sigma)$. If we denote by a_{ij} the coefficient of x_j in the i-th equation of this system, it will be verified without difficulty that the a_{ij} satisfy the relations

(13.4.2)

$$a_{i1}a_{j1} + a_{i2}a_{j2} + \cdots + a_{in}a_{jn} = \begin{cases} 1 & \text{for } i = j, \\ 0 & \text{for } i \neq j, \end{cases}$$

$$a_{1i}a_{1j} + a_{2i}a_{2j} + \cdots + a_{ni}a_{nj} = \begin{cases} 1 & \text{for } i = j, \\ 0 & \text{for } i \neq j. \end{cases}$$

These are straightforward generalizations of the well-known relations satisfied by the direction cosines of three mutually orthogonal vectors in three-dimensional space. Moreover, for $n = 3$ it is a familiar fact that a substitution of this type represents a *rotation* of a rectangular system of coordinates about the origin. By analogy we interpret the substitution 13.4.1 for any value of n as an *orthogonal substitution*, which represents a rotation about the origin of a rectangular system of coordinates in n-space, and we talk of the relations (13.4.2) satisfied by the coefficients as the *orthogonality relations*. The distance of any point from the origin must be invariant under a rotation of the coordinate system, and accordingly it follows from the orthogonality relations that we have

$$\sum_1^n x_i^2 = \sum_1^n y_i^2.$$

The determinant of the substitution 13.4.1 is easily calculated, and is found to have the value 1. Thus in particular it is different from zero, and by 10.4 the new variables $y_1, \ldots, y_n$ will then have a joint normal fr. f. From 13.4.1 and 13.4.2 we obtain

$$E(y_i) = 0, \qquad E(y_i^2) = \sigma^2, \qquad E(y_i\,y_j) = 0$$

for any $i \neq j$. Thus every y_i is normal $(0, \sigma)$, and the normally distributed variables $y_1, \cdots, y_n$ are uncorrelated, and hence by 10.4 also independent. This proves our assertion about the y_i.

It follows that the variables $y_1/\sigma, \cdots, y_n/\sigma$ are independent and normal $(0, 1)$.

186

Now we have, since $y_n = \bar{x}\sqrt{n}$,

$$n s^2 = \sum_1^n (x_i - \bar{x})^2$$

$$= \sum_1^n x_i^2 - n\bar{x}^2$$

$$= \sum_1^n y_i^2 - y_n^2$$

$$= \sum_1^{n-1} y_i^2,$$

and hence

$$\frac{n s^2}{\sigma^2} = \sum_1^{n-1} \left(\frac{y_i}{\sigma}\right)^2.$$

Thus the variable $n s^2/\sigma^2$ is equal to the sum of the squares of $n-1$ variables, which are independent and normal (0, 1). According to 8.1 this variable has a χ^2 distribution with $n-1$ degrees of freedom, the corresponding fr. f. being $k_{n-1}(x)$ as given by 8.1.1. Further, since all the variables $y_1, \ldots, y_{n-1}$ are independent of $y_n = \bar{x}\sqrt{n}$, it follows that $\bar{x}$ and s^2 are independent. — As already pointed out, all this remains true even if the assumption $m = 0$ is dropped, so that we may sum up our results in the following way.

In samples from a population which is normal (m, σ), the mean $\bar{x}$ and the variance s^2 are independent. $\bar{x}$ is normal $(m, \sigma/\sqrt{n})$, while $n s^2/\sigma^2$ has a χ^2 distribution with $n-1$ degrees of freedom.

Note that the variable $n s^2 = \sum_1^n (x_i - \bar{x})^2$ is primarily defined as the sum of n squares, not $n-1$. However, the squared variables are *not independent*, since they must satisfy the relation $\sum_1^n (x_i - \bar{x}) = 0$. By the substitution used above we have reduced $n s^2$ to a sum of *independent* or *free* squares. The constraint imposed upon the original variables by the relation they must satisfy then makes itself felt in the reduction of the number of squares from n to $n-1$. It then seems natural to introduce the expression *degrees of freedom*, and to say that the present problem involves $n-1$ degrees of freedom.

Consider now the two independent variables

$$Y = \sqrt{n}\,\frac{\bar{x} - m}{\sigma}, \quad Z = \frac{n s^2}{\sigma^2}.$$

Y is normal (0, 1), while Z has a χ^2 distribution with $n-1$ degrees of freedom. Let us write

187

$$(13.4.3) \qquad t = \sqrt{n-1}\, \frac{Y}{\sqrt{Z}} = \sqrt{n-1}\, \frac{\bar{x}-m}{s}.$$

From 8.2 we then obtain the important result that t has a STUDENT distribution with $n-1$ degrees of freedom, the corresponding fr. f. $s_{n-1}(x)$ being given by 8.2.1. It should be observed that the expression 13.4.3 of the "STUDENT ratio" t is independent of σ, while the fr. f. is independent of both m and σ. This will enable us to make an important use of t, in testing a hypothetical value of m, in a case when m and σ are both unknown (cf. 14.4).

The following is a further result in the same direction, which is also important for certain applications. Let $x_1, \ldots, x_{n_1}$ and $y_1, \ldots, y_{n_2}$ be observed values of the variables X and Y respectively, the latter being normal (m_1, σ) and (m_2, σ), so that the s. d. has the same value σ in both cases. Let $\bar{x} = \frac{1}{n_1}\sum x_i$ and $s_1^2 = \frac{1}{n_1}\sum(x_i-\bar{x})^2$ be the mean and the variance of the X sample, while $\bar{y}$ and s_2^2 are the corresponding characteristics of the Y sample. The variable $\frac{n_1 s_1^2 + n_2 s_2^2}{\sigma^2}$ will then (cf. 8.5, ex. 3) have a χ^2 distribution with $n_1 + n_2 - 2$ degrees of freedom, while according to 7.3 the variable $\bar{x} - \bar{y} - (m_1 - m_2)$ is normal $\left(0, \sigma\sqrt{\frac{1}{n_1} + \frac{1}{n_2}}\right)$. Writing

$$(13.4.4) \qquad t = \sqrt{\frac{n_1 n_2 (n_1 + n_2 - 2)}{n_1 + n_2}} \cdot \frac{\bar{x} - \bar{y} - (m_1 - m_2)}{\sqrt{n_1 s_1^2 + n_2 s_2^2}},$$

we can again use 8.2, and it is found that in this case t has a STUDENT distribution with $n_1 + n_2 - 2$ degrees of freedom. In certain cases, this result may be used to test whether the unknown means m_1 and m_2 of two samples may be assumed to be equal or not (cf. 14.4).

The STUDENT distribution also appears in connection with the sampling distributions of certain characteristics of two-dimensional samples (for detailed proofs, see *Mathematical Methods*, Ch. 29). Thus if X and Y have a normal joint fr. f., with the value β_{21} of the regression coefficient of Y on X, the variable

$$(13.4.5) \qquad t = \frac{s_1 \sqrt{n-2}}{s_2 \sqrt{1-r^2}} (b_{21} - \beta_{21})$$

has a STUDENT distribution with $n-2$ degrees of freedom. s_1, s_2, r and b_{21} are, of course, here the sample characteristics introduced in 12.4, for a sample of n observed pairs of values of the variables.

In the particular case when X and Y are independent, so that the correlation coefficient of the joint normal fr. f. is $\varrho = 0$, the variable

$$(13.4.6) \qquad t = \sqrt{n-2}\,\frac{r}{\sqrt{1-r^2}}$$

will also have a STUDENT distribution with $n-2$ degrees of freedom. In the general case, when $\varrho \neq 0$, the distribution of the sample correlation coefficient r is of a complicated nature, and cannot be reduced to STUDENT's distribution in the simple way shown by 13.4.6.

13.5. Problems. — 1. Find the sampling distribution of the mean $\bar{x}$ when the population has a) a CAUCHY distribution (cf. 5.9, ex. 2), b) a POISSON distribution, c) a χ^2 distribution.

2. Use the orthogonal transformation studied in 13.4, and also 8.5, ex. 4, to find the distribution of the variable $u = (x_n - \bar{x})/s$. Will this be different from the distribution of the variable $(x_i - \bar{x})/s$ for any $i \neq n$?

3. Using the notations of 13.4.4, find the distribution of the variable

$$v = \frac{a\,(\bar{x} - m_1) + b\,(\bar{y} - m_2)}{\sqrt{n_1\,s_1^2 + n_2\,s_2^2}},$$

where a and b are arbitrary constants.

4. With the same notations as in the preceding example, find the distribution of $w = s_1^2/s_2^2$.

5. A certain random variable X has a distribution of the continuous type. A sample of n values of X is given, and we denote these values, *arranged in order of increasing magnitude*, by $x_1, x_2, \ldots, x_n$. Then x_i is called the *i-th order statistic* of the sample. (In particular, when $n = 2m+1$ is odd, x_{m+1} will be the sample median.) Find the fr. f. of x_i.

6. Find the mean and the s. d. of the i-th order statistic x_i in the particular case when the variable has a rectangular distribution over the interval $(0, 1)$.

189

CHAPTER 14

PROBLEMS OF STATISTICAL INFERENCE

14.1. Introduction. — The concepts introduced in the preceding chapter will now be applied to various problems of statistical analysis and prediction, the general character of which has already been briefly indicated in Ch. 11.

We shall assume that we are dealing with a set of independently[1] observed values of certain random variables or, in the sampling terminology, with a sample of independently drawn values from some population having a certain distribution. Without actually knowing the distribution of the population, we are assumed to possess a certain amount of a priori information concerning its properties, and it is required to use the observed sample values in order to gain additional information in some specified sense.

Many different situations occurring in the applications are covered by this description of the typical problem of statistical inference. We shall here mainly consider the problems of *estimation* and of *testing hypotheses*, both of which have been briefly mentioned in 11.3. In both these cases, as well as in other questions of statistical inference, the approach to the problems will be highly dependent on the nature of the a priori information that we are assumed to possess about the distribution of the population.

In this respect, the case so far most thoroughly studied is when it can be assumed that the distribution has a known mathematical form, but contains a certain number of unknown parameters. It is then required to estimate the parameters, or to test some hypothesis bearing on their values. This is the *parametric case*, with which we shall be mainly concerned in the sequel. As an example, we may assume that a certain population has a normal distribution, with unknown values of m and σ, and we may then attempt to estimate m and σ by means of the sample values, or to test some hypothesis, such as the hypothesis that m has a given value m_0.

On the other hand, even with considerably less definite information—sometimes even without any information at all—about the distribution of the population, we may want to estimate some of its characteristics, such as the median or the interquartile range, or the value of the unknown d.f. $F(x)$ for some given x. Similarly we may want to test some hypothesis, such as the

---

[1] In 16.5 and 16.6 we shall consider some problems connected with sampling from a finite population, where the sample values are not independent. There are many further important problems involving samples of interdependent values, but most of them fall outside the scope of this book.

hypothesis that the median has a given value, etc. This is the *non-parametric case* of statistical inference, which is considerably less known than the former case, and will here only be mentioned in some simple examples.

In sections 14.2–14.4 we shall be concerned with various aspects of the estimation problem. We then proceed in 14.5 to some general remarks on the problem of testing hypotheses, while some simple examples will be given in 14.6–14.7. A particularly important case of testing hypotheses will be treated in Ch. 15, and finally various further examples of statistical inference are given in Ch. 16.

14.2. Properties of estimates. — We shall now consider the parametric case of estimation, but the main concepts introduced in the present section will apply also in the non-parametric case. — We begin with the simple case of one single unknown parameter. Let $F(x; \alpha)$ be a d.f. of known mathematical form, containing an unspecified constant parameter α. We assume that our sample values $x_1, \ldots, x_n$ have been drawn from a population with the d.f. $F(x; \alpha)$, where the parameter has a certain constant, but unknown value, and it is required to use the sample values in order to estimate α.

Thus we want to form some function $\alpha^* = \alpha^* (x_1, \ldots, x_n)$ of the sample values, not containing the unknown parameter α, and such that α^* may be used as a reasonable estimate of the unknown value α. Naturally we should like this to be a *good* estimate, and so the question arises: what do we mean by a good estimate?

Any function $\alpha^*(x_1, \ldots, x_n)$ will be a random variable, having a certain probability distribution, Since every x_i has the d.f. $F(x; \alpha)$ depending on the unknown α, the distribution of $\alpha^*(x_1, \ldots, x_n)$ will as a rule itself depend on α. The estimate obtained from a particular set of observed sample values is an observed value of this variable, and we want to be able to assert that with a large probability this estimate will fall near the true value of α. Obviously this requires that *the probability distribution of α^* should be concentrated as closely as possible near the true parameter value* α. From another point of view the same thing may be expressed by saying that the dispersion of the α^* distribution about the point α should be as small as possible, — This is the fundamental condition to be satisfied by a good estimate. The condition is somewhat vague, and can be interpreted in different ways, thus giving rise to various classes of "good" estimates. In this respect, several concepts are currently used in the literature, some of which will now be defined.[1]

---

[1] For proofs and further developments of the properties of estimates discussed in this section, the reader is referred to *Mathematical Methods*, Chapter 32.

An estimate $\alpha^* = \alpha^*(x_1, \ldots, x_n)$ is called *unbiased* if we have

$$E(\alpha^*) = \alpha.$$

Thus it follows from 13.2.1 that the sample mean $\bar{x}$ is an unbiased estimate of the population mean m, while 13.2.2 shows that the sample variance s^2, as estimate of σ^2, has a *negative bias*, $E(s^2)$ being always smaller than σ^2. However, the difference is small for large values of n. Moreover, as pointed out in 13.2, the estimate s^2 may be corrected for bias by multiplication with the factor $\dfrac{n}{n-1}$. This is a typical situation. When estimates are biased, it is often possible to find some simple correction that removes the bias altogether. Even when this cannot be done, a biased estimate can be useful, provided only that we can show that the amount of the bias is small, which is often the case in large samples. On the other hand, even an unbiased estimate will be of little use as long as we do not know the degree of dispersion of its probability distribution about the true value α.

The dispersion of a probability distribution about a given point can be measured in several different ways (cf. 5.7). A natural measure is provided by the second order moment about the given point, and accordingly the *efficiency* of different possible estimates of the same parameter α, say α_1^* and α_2^*, can be compared by means of the corresponding mean values $E(\alpha_1^* - \alpha)^2$ and $E(\alpha_2^* - \alpha)^2$. Clearly the estimate having the smallest $E(\alpha^* - \alpha)^2$ is the more efficient one. Accordingly the *relative efficiency* of α_2^*, as compared with α_1^*, is defined as the ratio

$$\frac{E\,(\alpha_1^* - \alpha)^2}{E\,(\alpha_2^* - \alpha)^2}.$$

If this ratio is > 1, α_2^* is more efficient than α_1^*, and vice versa. — For an unbiased estimate α^*, the mean value $E(\alpha^* - \alpha)^2$ is, of course, identical with the variance of α^*.

Sometimes we are particularly interested in the properties of our estimates in *large samples*, i.e., for large values of n. Let $\alpha^* = \alpha^*(x_1, \ldots, x_n)$ be an estimate which is defined for arbitrary large values of n. If the mean value $E(\alpha^* - \alpha)^2$ tends to zero as $n \to \infty$, this implies by TCHEBYCHEFF's theorem (cf. 5.8) that for large n the mass in the α^* distribution is concentrated near the point α. Consequently, for any given $\varepsilon > 0$ the probability $P(|\alpha^* - \alpha| < \varepsilon)$ that the estimate α^* will differ by less than ε from the true value α tends to unity as $n \to \infty$. When this holds, we shall say that α^* is a *consistent* estimate of α.

In the majority of cases, our estimates will be functions of some of the ordinary sample characteristics discussed in Chs. 12 and 13. As pointed out in 13.2 and 13.3, such an estimate α^* will usually be asymptotically normally distributed, and the corresponding variance $D^2(\alpha^*)$ will be small for large n, with a dominating term of the form c^2/n, where c is a constant depending on the nature of the estimate. Suppose, for the sake of simplicity, that we are dealing with an unbiased estimate. The quantity $\sqrt{n}(\alpha^* - \alpha)$ will then be asymptotically normal $(0, c)$. (This is, of course, a much stronger property than consistency.) In some cases it is possible to show that, among all estimates having these properties, a certain α_0^* has the smallest possible value of the constant c. Then α_0^* is said to be an *asymptotically efficient* estimate.

All these concepts are immediately extended to samples from multi-dimensional distributions, and to cases of more than one unknown parameter. Thus in the case of a two-dimensional sample consisting of n pairs of values (x_1, y_1), $\ldots$, (x_n, y_n), an estimate of the unknown parameter α will in general be a function $\alpha^* = \alpha^*(x_1, \ldots, x_n, y_1, \ldots, y_n)$ depending on all $2n$ sample values, while in the case of two unknown parameters α and β we shall have to consider estimates α^* and β^* for both parameters. With these evident modifications, all definitions given above still hold. In the case of several unknown parameters, we can also define a *joint efficiency* of our estimates for all parameters, but this concept is somewhat more complicated and will not be used here.

Ex. — According to 13.2.1, the sample mean $\bar{x}$ is for any distribution[1] an unbiased and consistent estimate of the population mean m.

Let now $f(x)$ be a given fr.f. with the mean zero and the known s.d. σ, and such that $f(x) = f(-x)$. Suppose that we have a sample of n values from a population with the fr.f. $f(x - m)$, where m is unknown. Since the mean and the median of the population are both equal to m, we can use the mean $\bar{x}$ and the median z_2 of the sample as alternative estimates of m. Both estimates are asymptotically normal (cf. 13.3), their variances in large samples being asymptotically σ^2/n for the mean, and $1/(4nf^2(0))$ for the median. Hence the relative efficiency of the median, as compared with the mean, is for large n approximately $4\sigma^2 f^2(0)$.

Consider first the case when $f(x)$ is a normal fr.f., $f(x) = \dfrac{1}{\sigma\sqrt{2\pi}} e^{-\frac{x^2}{2}}$. In this case the relative efficiency of the median becomes $\dfrac{2}{\pi} = 0.637$. In samples from a normal distribution, the mean will thus provide a better estimate of m than the median. In fact, it can be proved (cf. *Mathematical Methods*, p. 483) that among all unbiased estimates of m in normal samples of given size n, the mean $\bar{x}$ hae the smallest possible

---

[1] Here, as elsewhere in this chapter, we have tacitly assumed that σ is finite. If this assumption is dropped, $\bar{x}$ will not necessarily be a consistent estimate, as shown by the example of the CAUCHY distribution (cf. 5.9, Ex. 2, and the concluding remark of 7.3).

variance. Thus for normal samples $\bar{x}$ is not only asymptotically efficient, but has even the stronger property of minimizing the variance for any fixed sample size n.

Take, on the other hand, $f(x) = s_k(x)$, the fr.f. of STUDENT's distribution with k degrees of freedom, and suppose $k \geq 3$, so that the first and second order moments are both finite. From 8.2 we then obtain for the relative efficiency of the median

$$4\,\sigma^2 f^2(0) = \frac{4}{(k-2)\,\pi}\left(\frac{\Gamma\left(\dfrac{k+1}{2}\right)}{\Gamma\left(\dfrac{k}{2}\right)}\right)^2.$$

A simple calculation will show that for $k = 3$ this expression takes the value 1.621, and for $k = 4$ the value 1.125. Thus for these values of k the relative efficiency of the median is > 1, so that the median gives a better estimate than the mean. For $k \geq 5$, on the other hand, the relative efficiency of the median becomes < 1, which shows the superiority of the mean for these values of k. As $k \to \infty$ the relative efficiency tends to the value $2/\pi$ obtained above for the normal distribution.

14.3. Methods of estimation. — The most important general method of finding estimates is the *maximum likelihood method* introduced by R. A. FISHER. We shall give an account of this method for the simple case of a sample drawn from a one-dimensional distribution with one single unknown parameter. The modifications required in more general cases will be easily found. The method takes a slightly different form according as the population has a distribution of the continuous or of the discrete type.

For a sample of n values $x_1, \ldots, x_n$ drawn from a distribution of the *continuous* type, with the fr.f. $f(x; \alpha)$, we define the *likelihood function* $L(x_1, \ldots, x_n; \alpha)$ by the relation

$$(14.3.1) \qquad L(x_1, \ldots, x_n; \alpha) = f(x_1; \alpha) \cdots f(x_n; \alpha).$$

Suppose, on the other hand, that our sample has been drawn from a *discrete* distribution, where the possible values of the variable are $\xi_1, \ldots, \xi_r$ with the corresponding observed frequencies $f_1, \ldots, f_r$, so that $f_1 + \cdots + f_r = n$. We restrict ourselves to the case when the ξ_i are independent of the parameter α, while the corresponding probabilities may depend on α:

$$P(X = \xi_i) = p_i(\alpha).$$

In this case we define the likelihood function by writing

$$(14.3.2) \qquad L(x_1, \ldots, x_n; \alpha) = (p_1(\alpha))^{f_1} \cdots (p_r(\alpha))^{f_r}.$$

When the sample values are given, the likelihood function is in either case a function of the parameter α. *The maximum likelihood method consists in choosing, as an estimate of the unknown population value of α, the value that renders the likelihood function as large as possible.*

Since $\log L$ attains its maximum for the same value of α as L, the well-known elementary rules for finding maxima and minima will lead us to consider the equation

$$(14.3.3) \qquad \frac{\partial \log L}{\partial \alpha} = 0.$$

In some cases we may be able to solve this *likelihood equation* explicitly with respect to α, while in other cases we shall have to be satisfied with a numerical determination of the root or roots corresponding to a given set of sample values. Any root of 14.3.3 will be a function of the sample values

$$\alpha^* = \alpha^* (x_1, \ldots, x_n),$$

and if the root corresponds to a maximum of L, it will be denoted as a *maximum likelihood estimate of α*.

It can be shown that, under fairly general conditions, the maximum likelihood method will yield asymptotically efficient estimates, as defined in 14.2. For the proof of this, and other important properties of the method, we refer to *Mathematical Methods*, Ch. 33.

In the case of a distribution containing two or more unknown parameters, there will be one likelihood equation for each parameter. Thus, e. g., for two unknown parameters α and β, the likelihood function L will contain both parameters, and the maximum condition will lead to the equations

$$\frac{\partial \log L}{\partial \alpha} = 0 \text{ and } \frac{\partial \log L}{\partial \beta} = 0.$$

Any pair of roots of these equations corresponding to a maximum of L will constitute a pair of joint maximum likelihood estimates of α and β.

The maximum likelihood method may be used for the estimation of parameters under much more general conditions than those considered here. Suppose, e.g., that $x_1, \ldots, x_n$ are random variables having a joint fr.f. $f(x_1, \ldots, x_n; \alpha)$ containing the unknown parameter α. We observe one system of values of all n variables, and choose as our estimate of α the value that renders the function f as large as possible for the observed values of x_i. Obviously this is a generalized form of the method given above for a continuous distribution, which corresponds to the particular case when the x_i are independent variables all having the same fr.f. $f(x; \alpha)$. A similar generalization can be made in the discrete case.

A further general estimation method is the *method of moments* developed by K. PEARSON. This method consists in equating a convenient number of the sample moments to the corresponding moments of the distribution, which are functions of the unknown parameters. By considering as many moments as there are parameters to be estimated, and solving the resulting equations with respect to the parameters, estimates of the latter are obtained. This method will often lead to simpler calculations than the maximum likelihood method, but on the other hand the estimations obtained are usually less efficient. Nevertheless, on account of its practical expediency, the method of moments, and even still less efficient methods, will often render good service.

Ex. 1. — Suppose that, in connection with a certain random experiment, the event A has a constant, but unknown probability p. In a series of n repetitions of the experiment, we observe f times the occurrence of A. It is required to estimate the probability p.

If we define a random variable X by taking $X = 1$ when A occurs, and otherwise $X = 0$, this variable will have a two point distribution defined by

$$P(X = 1) = p, \qquad P(X = 0) = 1 - p,$$

where p is an unknown parameter. From our observations we have obtained a sample of n observed values of the variable, f of which are $= 1$, while $n - f$ are $= 0$. From 14.3.2 we then obtain the likelihood function

$$L = p^f (1 - p)^{n-f},$$

and the maximum likelihood method leads to the equation

$$\frac{\partial \log L}{\partial p} = \frac{f}{p} - \frac{n - f}{1 - p} = 0,$$

with the unique solution $p = f/n$. We thus arrive at the fairly obvious result that the observed frequency ratio f/n should be taken as an estimate of p. We already know (cf. 6.3 and 6.4) that this is an unbiased, consistent and asymptotically normal estimate. From the general properties of maximum likelihood estimates it follows that it is asymptotically efficient, and it can be shown (cf. *Mathematical Methods*, p. 487) that it even has the stronger property of minimizing the variance for any fixed sample size n.

Ex. 2. — Suppose that the variable X has a POISSON distribution with an unknown parameter λ. It is required to estimate λ by means of a sample of n observed values $x_1, \ldots, x_n$.

In this case the variable has an infinite sequence of possible values: 0, 1, 2, ... However, only a finite number of these values can be represented in our sample. Let r denote the largest value of X observed in our sample, and let the values 0, 1, ..., r have the frequencies $f_0, f_1, \ldots, f_r$, so that $f_0 + \ldots + f_r = n$. The probability that X takes the value i is now (cf. 6.5)

$$p_i(\lambda) = \frac{\lambda^i}{i!} e^{-\lambda},$$

and the likelihood function becomes

$$L = \prod_0^r \left(\frac{\lambda^i}{i!} e^{-\lambda}\right)^{f_i}.$$

We thus obtain the equation

$$\frac{\partial \log L}{\partial \lambda} = \sum_0^r f_i \left(\frac{i}{\lambda} - 1\right) = 0$$

which gives

$$\lambda = \frac{\sum\limits_0^r i f_i}{\sum\limits_0^r f_i} = \frac{1}{n} \sum_1^n x_i = \bar{x},$$

since we have $\sum\limits_0^r i f_i = \sum\limits_1^n x_i$.

The maximum likelihood estimate of λ will thus simply be the sample mean $\bar{x}$. Thus if we assume that the blood corpuscle data of Table 12.1.1 are sample values from a POISSON distribution, the numerical computation of Table 12.5.1 gives the estimated value $\bar{x} = 11.91$ for the parameter λ. — Since the mean of the distribution is $m = \lambda$, the estimate is unbiased.

Ex. 3. — Suppose now that the variable X is normal (m, σ), where both parameters m and σ are unknown. According to 14.3.1 the likelihood function of a sample consisting of the observed values $x_1, \ldots, x_n$ is

$$L = \prod_1^n \frac{1}{\sigma \sqrt{2\pi}} e^{-\frac{(x_i - m)^2}{2\sigma^2}},$$

and the maximum likelihood method gives the equations

$$\frac{\partial \log L}{\partial m} = \sum_1^n \frac{x_i - m}{\sigma^2} = 0,$$

$$\frac{\partial \log L}{\partial \sigma} = \sum_1^n \left(-\frac{1}{\sigma} + \frac{(x_i - m)^2}{\sigma^3}\right) = 0,$$

from which we obtain

$$m = \frac{1}{n} \sum_1^n x_i = \bar{x},$$

$$\sigma^2 = \frac{1}{n} \sum_1^n (x_i - \bar{x})^2 = s^2.$$

197

Thus, in the normal case, the maximum likelihood estimates of the mean and the variance of the distribution are the corresponding characteristics of the sample. We already know that $\bar{x}$ gives an unbiased estimate of m, while s^2 is biased, $E(s^2)$ being equal to $\frac{n-1}{n}\sigma^2$. However, the factor $\frac{n-1}{n}$ differs very little from 1 in large samples, and this bias does not prevent s^2 from being, like $\bar{x}$, an asymptotically efficient estimate. (Note that $\bar{x}$ has, as mentioned in 14.2, even the stronger property of minimizing the variance for any fixed sample size n.) — For the s.d. σ we obtain from the above equation the maximum likelihood estimate s, which is also asymptotically efficient, though slightly biased.

There are, of course, many other ways of estimating m and σ from a normal sample. In 14.2 we have seen that the sample median gives an estimate of m having a relative efficiency of about 64% as compared with the mean. Other possible estimates of m are e.g. the arithmetic mean of the first and third quartiles, $\frac{1}{2}(z_1 + z_3)$, and the mean of the extreme sample values, $\frac{1}{2}(x_{max} + x_{min})$. All such estimates have a lower efficiency than the mean $\bar{x}$. The s.d. σ may also be estimated by means of the extreme sample values. In fact, it is easily seen that the mean value of the range of a normal sample is $E(x_{max} - x_{min}) = c_n\,\sigma$, where c_n is a constant depending only on n. Thus the quantity $(x_{max} - x_{min})/c_n$ gives an unbiased estimate of σ. Though this is an estimate of considerably lower efficiency than s, it will sometimes be of practical use, since it can be computed extremely simply, the factors c_n being available in existing tables.

Ex. 4. — Consider finally the case of a variable X uniformly distributed (cf. 5.3, Ex. 1) over the interval $(0, \alpha)$, where α is an unknown parameter. The fr.f. $f(x; \alpha)$ is equal to $1/\alpha$ for $0 < x < \alpha$, and otherwise equal to zero. Since all our sample values must lie between 0 and α, the likelihood function will be $L = 1/\alpha^n$. In order to render L as large as possible, we must take α as small as possible. In this case the value of α for which the maximum is attained cannot be found by differentiation, but it is immediately seen that the smallest value of α compatible with the given conditions is obtained by taking $\alpha = x_{max}$, the largest among the sample values. Thus x_{max} will be the maximum likelihood estimate of α. It can be shown without difficulty that this is a consistent estimate having a small negative bias, $E(x_{max})$ being equal to $\frac{n}{n+1}\alpha$ (cf. 13.5, ex. 6).

14.4. Confidence intervals, I.

— In the theory of estimation developed in the two preceding sections, any unknown parameter is estimated by one single quantity. An estimation of this kind is usually not sufficient to provide a satisfactory solution of a practical problem. If we are told, e.g., that from statistical observations the probability of recovery from a certain illness is estimated at 70% for cases treated by method A, and 80% for cases treated by method B, we shall probably feel that this is very well, but that we should like to have some idea of the *precision* of these estimates, before they can be confidently used as a basis for practical action.

In order to obtain a measure of the precision of an estimate α^* of the unknown parameter α, we might try to find two positive numbers δ and ε such that the probability that the true parameter value α is included between the limits $\alpha^* \pm \delta$ is equal to $1 - \varepsilon$:

(14.4.1) $$P\left(\alpha^* - \delta < \alpha < \alpha^* + \delta\right) = 1 - \varepsilon.$$

For a given probability $1 - \varepsilon$, high precision of the estimate would then obviously be associated with small values of δ.

14.4.1 expresses that the probability that the interval $(\alpha^* - \delta, \alpha^* + \delta)$ includes the true value α is equal to $1 - \varepsilon$. More generally, we may ask if it would be possible to work out a theory of *estimation by interval*, as distinct from the *point estimation* studied in the preceding sections. To an unknown parameter α, we should then try to find two functions of the sample values, α_1^* and α_2^*, such that the probability that the interval (α_1^*, α_2^*) includes the true parameter value α has a given value $1 - \varepsilon$:

(14.4.2) $$P\left(\alpha_1^* < \alpha < \alpha_2^*\right) = 1 - \varepsilon.$$

Such an interval, when it exists, will be called a *confidence interval* for the parameter α, and the probability $1 - \varepsilon$ will be denoted as the *confidence coefficient* of the interval. These concepts are due to J. NEYMAN.

In the present section, we shall confine ourselves to *samples from normal distributions*, and we shall show how in this case confidence intervals for various parameters can be obtained in a simple way, and how they can be used for the solution of practical problems. In the following section we shall discuss the corresponding problems for more general distributions, where in most cases we shall have to be content with approximate solutions.

1. *The mean m.* — Let the variable X be normal (m, σ), where m and σ are both unknown. Suppose that we have a sample of n observed values x_1, $\ldots, x_n$, and that it is required to find a confidence interval for the mean m. By 13.4.3 we know that the variable

(14.4.3) $$t = \sqrt{n-1} \, \frac{\bar{x} - m}{s}$$

has a STUDENT distribution with $n-1$ degrees of freedom. Let now p denote a given percentage, and let t_p be the p percent value of t for $n-1$ degrees of freedom (cf. 8.2), which can be directly found from Table IV. By the definition of t_p we then have

(14.4.4) $$P\left(-t_p < \sqrt{n-1} \, \frac{\bar{x} - m}{s} < t_p\right) = 1 - \frac{p}{100}.$$

However, it is easily seen that the double inequality

$$-t_p < \sqrt{n-1}\,\frac{\bar{x}-m}{s} < t_p$$

can be brought by simple transformations into the completely equivalent form

$$\bar{x} - t_p\,\frac{s}{\sqrt{n-1}} < m < \bar{x} + t_p\,\frac{s}{\sqrt{n-1}}\,.$$

Thus 14.4.4 can also be written

$$(14.4.5) \qquad P\left(\bar{x} - t_p\,\frac{s}{\sqrt{n-1}} < m < \bar{x} + t_p\,\frac{s}{\sqrt{n-1}}\right) = 1 - \frac{p}{100}\,.$$

The two relations 14.4.4 and 14.4.5, though completely equivalent, differ in one important respect. 14.4.4 states that we have the probability $100 - p$ % that the random variable $t = \sqrt{n-1}\,\dfrac{\bar{x}-m}{s}$ will take a value between the two numerically fixed limits $\pm t_p$. On the other hand, 14.4.5 states that we have the same probability $100 - p$ % that the two random variables $\bar{x} \pm t_p\,\dfrac{s}{\sqrt{n-1}}$ will takes values including between them the fixed, but unknown true value of the parameter m. Obviously 14.4.5 is a relation of the type tentatively suggested in 14.4.2. Accordingly the interval

$$\left(\bar{x} - t_p\,\frac{s}{\sqrt{n-1}},\ \bar{x} + t_p\,\frac{s}{\sqrt{n-1}}\right)$$

is a $100 - p$ % *confidence interval for* m, the limits of the interval are *confidence limits for* m, and the corresponding *confidence coefficient* is $1 - \dfrac{p}{100}\,.$

These concepts should be interpreted in the following way. Let us first choose once for all a fixed percentage p. Consider then a sequence of independent experiments, where each experiment consists in drawing a sample of n values from a population which is normal $(m,\ \sigma)$, the sample size n and the parameters m and σ being at liberty kept constant or allowed to vary from experiment to experiment in a perfectly arbitrary way, random or non-random. For each sample we compute the confidence limits $\bar{x} \pm t_p\,\dfrac{s}{\sqrt{n-1}}$, *and then we state that the mean* m *of the corresponding population is situated between these limits.* The confidence limits will obviously vary from sample to sample. However, it follows from 14.4.5 that the probability that our statement will be true is constantly equal to $100 - p$ %, and consequently the probability of making a false statement will always be p %. In the long run we may thus expect our statements to be false in about p % of all cases, and otherwise correct.

We thus find that the confidence interval specified by 14.4.5 provides us with a rule for estimating the parameter m, which is associated with a constant risk of error equal to p %, where p can be arbitrarily chosen.

The constant percentage p is known as the *level of significance*, or briefly the *level*, of the rule of estimation. If we want to work with a very small risk of error, we shall obviously have to choose a small value of p. However, when p tends to zero, the p percent value t_p will tend to infinity. Since the length of the confidence interval is proportional to t_p, it follows that a small level will be associated with a long confidence interval, i.e., with a low precision in the estimation of m, and vice versa. In practice, the choice of the level should thus be the result of a compromise between the opposite claims for low risk and high precision. In many fields of application it has become customary to work preferably on one of the levels 5%, 1%, or 0.1%.

It is easily seen that the confidence interval given by 14.4.5 is by no means unique. In 14.4.4 the limits $-t_p$ and $+t_p$ can be replaced by $-t'$ and $+t''$ respectively, where t' and t'' are any numbers such that

$$\int_{-t'}^{t''} s_{n-1}(t)\, dt = 1 - \frac{p}{100},$$

s_{n-1} being the fr.f. of STUDENT's distribution with $n-1$ degrees of freedom. The same argument as above will then lead to the confidence interval

$$\left(\bar{x} - t'' \frac{s}{\sqrt{n-1}}, \quad \bar{x} + t' \frac{s}{\sqrt{n-1}} \right),$$

which is still associated with the same confidence coefficient $1 - \frac{p}{100}$. It will be readily seen that among all confidence intervals obtained in this way for a given value of p, the one corresponding to $t' = t'' = t_p$ is *shorter* than any other. Thus the confidence interval given by 14.4.5 provides, among all these, the highest precision compatible with a given level of significance.

In certain applications we are interested in testing whether our observed sample agrees with the hypothesis that m has some given value, say $m = m_0$. A simple way of testing this hypothesis offers itself naturally in connection with the confidence interval deduced above. Working on a given level $p\%$, we first compute the confidence limits of m according to 14.4.5. If the given point m_0 falls outside the confidence interval, we say that m *differs significantly from m_0 on the $p\%$ level*, and accordingly we *reject* the hypothesis $m = m_0$. On the other hand, if the confidence interval includes the point m_0, we say that *no significant difference has been found*, and accordingly we *accept* the hypothesis $m = m_0$.

In a case when the hypothesis $m = m_0$ is in fact true, it follows from 14.4.5 that this test gives us the probability $1 - \dfrac{p}{100}$ of accepting the hypothesis, and consequently the probability $\dfrac{p}{100}$ of rejecting. The probability of committing an error by rejecting the hypothesis when it is in fact true is thus equal to the *level of the test, $p\%$*.

In order to apply this test to a particular sample, we only have to find the sample characteristics $\bar{x}$ and s, and from them compute the value

$$(14.4.6) \qquad\qquad t = \sqrt{n-1}\,\frac{\bar{x} - m_0}{s}.$$

Denoting as usual by p the desired level of the test, we then look up the value t_p in the row corresponding to $n-1$ degrees of freedom in Table IV. If $|t| > t_p$, the difference is significant on the level $p\%$, and vice versa.

In the numerical examples given below, a value of t found in this way will be denoted as *almost significant* if $|t|$ exceeds the 5% value, but falls short of the 1% value. Further, the t value will be called *significant* if $|t|$ lies between the 1% and the 0.1% values, and *highly significant* if $|t|$ exceeds the 0.1% value. This is, of course, a purely conventional terminology. We shall use corresponding expressions in respect of other similar tests.

Ex. 1. — Assume that the length of life of an electric bulb made by a given manufacturing procedure is a normally distributed variable. For a sample of $n = 20$ bulbs the lengths of life have been determined by actual observation, and the sample characteristics $\bar{x} = 1832$ hours and $s = 497$ hours have been found. For $n-1 = 19$ degrees of freedom we then obtain the values of t_p and the confidence limits for the unknown mean m of the length of life of a bulb, which are given below:

Level of significance	$p \%$ value for 19 d. of fr.	Confidence limits	
		Lower	Upper
p	t_p	$\bar{x} - t_p \dfrac{s}{\sqrt{n-1}}$	$\bar{x} + t_p \dfrac{s}{\sqrt{n-1}}$
5 %	2.093	1593	2071
1 %	2.861	1506	2158
0.1 %	3.883	1389	2275

Thus e.g. the statement that the point m is included in the confidence interval with the limits 1593 and 2071 has a confidence coefficient of 95%. It is essential to be quite clear about the meaning of this statement. We are *not* saying that a certain variable m has the probability 95% of falling between the given limits 1593 and 2071. *The mean m is not a random variable, but a fixed, though unknown constant.* As already pointed out in connection with 14.4.5 it is the *confidence limits* that are random variables, and on the level 5% we have the probability 95% that these limits will take values including between them the constant m. In our case we have observed one value of each of these variables, and the observed values are 1593 and 2071.

Suppose, on the other hand, that we are not interested in an estimation of m, but just want to find out whether our observed sample agrees with the hypothesis that m has some a priori given value, say $m_0 = 2000$ hours. We then have to compute t from 14.4.6, which gives $t = -1.473$. Since $|t|$ does not even exceed the 5% value for 19 degrees of freedom, which is 2.093, the difference is not even almost significant. Thus the hypothesis $m = 2000$ should be accepted, at least until further observations are available.

2. *The standard deviation σ.* — Still supposing that we have a sample of n values from a normal distribution with unknown values of m and σ, we shall now consider the estimation of σ. The argument is perfectly analogous to the preceding case. By 13.4 the variable $n s^2/\sigma^2$ has a χ^2 distribution with $n-1$ degrees of freedom. For any given level $p \%$, we can find infinitely many intervals, each of which contains exactly the mass $1 - \dfrac{p}{100}$ in this distribution. Among all these intervals, we shall choose the particular interval $(\chi^2_{p'}, \chi^2_{p''})$, where $\chi^2_{p'}$ and $\chi^2_{p''}$ are the p' and p'' percent values of the χ^2 distribution for $n-1$ degrees of freedom (cf. 8.1), while

$$p' = 100 - \tfrac{1}{2} p, \quad p'' = \tfrac{1}{2} p.$$

Each of the tails $\chi^2 < \chi^2_{p'}$ and $\chi^2 > \chi^2_{p''}$ will then contain the mass $\tfrac{1}{2} p \%$, and we have

(14.4.7) $$P\left(\chi^2_{p'} < \frac{n s^2}{\sigma^2} < \chi^2_{p''}\right) = 1 - \frac{p}{100}.$$

203

By a simple transformation of the double inequality within the brackets we obtain, in the same way as 14.4.5 was obtained from 14.4.4, the equivalent relation

$$(14.4.8) \qquad P\left(\frac{n s^2}{\chi_{p''}^2} < \sigma^2 < \frac{n s^2}{\chi_{p'}^2}\right) = 1 - \frac{p}{100},$$

which can also be written

$$(14.4.9) \qquad P\left(\frac{s \sqrt{n}}{\chi_{p''}} < \sigma < \frac{s \sqrt{n}}{\chi_{p'}}\right) = 1 - \frac{p}{100}.$$

The comparison between 14.4.7 and 14.4.8 gives rise to the same remarks as the comparison between 14.4.4 and 14.4.5. For the variance σ^2, we here obtain the confidence limits $\frac{n s^2}{\chi_{p''}^2}$ and $\frac{n s^2}{\chi_{p'}^2}$, corresponding to the level p %. For the standard deviation σ, 14.4.9 gives the confidence limits $\frac{s \sqrt{n}}{\chi_{p''}}$ and $\frac{s \sqrt{n}}{\chi_{p'}}$. — By means of the confidence limits thus obtained, the hypothesis that σ has some given value σ_0 can be tested in a similar way as explained above in the case of m.

Ex. 2. — In the case of the 20 electric bulbs considered in ex. 1 we find for the level $p = 5$ % by interpolation in the row for 19 degrees of freedom in Table III

$$\chi_{p'} = \chi_{97.5}^2 = 8.825, \quad \chi_{p''} = \chi_{2.5}^2 = 33.096.$$

The confidence limits for σ then become $\frac{s \sqrt{n}}{\chi_{p''}} = 386$ and $\frac{s \sqrt{n}}{\chi_{p'}} = 748$.

3. *The difference between two mean values.* — A case which frequently comes up in the applications is the following. Two independently observed samples from normally distributed populations are available, and it is required to decide whether the unknown means of the two populations are equal or unequal. Denoting the two means by m_1 and m_2, we are then concerned with the hypothesis that $m_1 = m_2$, and we have to decide whether this hypothesis should, on the basis of our data, be accepted or rejected.

In some cases of this type it can be a priori assumed that the variances of the two populations, though unknown, are equal. We shall here only treat the problem under this simplifying assumption, the general case being somewhat intricate.

Suppose, as in 13.4, that the two samples contain n_1 and n_2 observations respectively, and let the sample means be denoted by $\bar{x}$ and $\bar{y}$, while s_1^2 and s_2^2 are the sample variances. According to 13.4.4, the variable

$$(14.4.10) \qquad t = \sqrt{\frac{n_1 n_2 (n_1 + n_2 - 2)}{n_1 + n_2}} \cdot \frac{\bar{x} - \bar{y} - (m_1 - m_2)}{\sqrt{n_1 s_1^2 + n_2 s_2^2}}$$

has a STUDENT distribution with $n_1 + n_2 - 2$ degrees of freedom. By an easy adaptation of the argument used in the estimation of m, we then obtain for the difference $m_1 - m_2$ between the unknown true mean values the confidence limits

$$(14.4.11) \qquad \bar{x} - \bar{y} \pm t_p \sqrt{\frac{(n_1 + n_2)(n_1 s_1^2 + n_2 s_2^2)}{n_1 n_2 (n_1 + n_2 - 2)}},$$

where t_p is the $p \%$ value of t taken from the row for $n_1 + n_2 - 2$ degrees of freedom in Table IV, the level of significance being as usual $p \%$.

In order to test the hypothesis $m_1 = m_2$, or $m_1 - m_2 = 0$, we are now as in the case of m led to the following simple test. After having computed the confidence limits given by 14.4.11, we see whether the confidence interval thus obtained includes the point 0 or not. In the former case the hypothesis may be accepted, while in the latter case it should be rejected.

In a case when the hypothesis $m_1 - m_2$ is in fact true, we have according to the definition of the confidence interval the probability $p \%$ that the confidence interval does *not* include the point 0. Thus in adopting the above test we shall as in the case of m have the prescribed probability $p \%$ of committing an error by rejecting the hypothesis in a case when it is in fact true.

In order to apply this test in practice, we have only to substitute in 14.4.10 the hypothetical value $m_1 - m_2 = 0$, and compute the value

$$(14.4.12) \qquad t = (\bar{x} - \bar{y}) \sqrt{\frac{n_1 n_2 (n_1 + n_2 - 2)}{(n_1 + n_2)(n_1 s_1^2 + n_2 s_2^2)}}.$$

Looking up the value t_p corresponding to $n_1 + n_2 - 2$ degrees of freedom, we then see whether $|t| > t_p$ or not. In the former case, we say that m_1 and m_2 differ significantly on the $p \%$ level, while in the latter case we conclude that no significant difference has been found in the observed samples, so that the hypothesis $m_1 = m_2$ can be accepted, at least until further observations are available.

Ex. 3. — Suppose that, in addition to the sample of $n_1 = 20$ electric bulbs considered in Exs. 1 and 2, we have observed a further sample of $n_2 = 30$ bulbs manufactured according to a different method, and that we have found the sample characteristics

$$n_1 = 20, \quad \bar{x} = 1832, \quad s_1 = 497,$$

$$n_2 = 30, \quad \bar{y} = 1261, \quad s_2 = 501.$$

Suppose further that the change of method can be reasonably assumed not to affect the variance of the length of life of a bulb, while the mean may possibly be affected. In order to test the hypothesis that the means m_1 and m_2 of the two populations are equal, we then compute the value t from 14.4.12, and find $t = 3.881$. This exceeds even the 0.1% value for $n_1 + n_2 - 2 = 48$ degrees of freedom, which by interpolation in Table IV is found to be 3.515. According to the conventional terminology introduced above in the case of m, we shall thus say that there is a "highly significant" difference between the means, and that the hypothesis $m_1 = m_2$ should be rejected.

On the level $p = 0.1\%$, the confidence limits for the difference $m_1 - m_2$ computed from 14.4.11 are 54 and 1088. If we make the statement that, in respect of the mean length of life of a bulb, method 1 exceeds method 2 by at least 54 hours, this will be a statement made in accordance with a rule affected by a probability of error of at most 0.1%.

4. *The regression coefficient.* — Suppose that the variables X and Y have a joint normal distribution, and let β_{21} be the regression coefficient of Y on X in this distribution. If b_{21} is the corresponding observed regression coefficient, calculated from a sample of n observed pairs of values of the variables, we have seen in 13.4.5 that the variable

$$(14.4.13) \qquad t = \frac{s_1 \sqrt{n-2}}{s_2 \sqrt{1-r^2}} (b_{21} - \beta_{21})$$

will have a STUDENT distribution with $n - 2$ degrees of freedom. In the same way as above, we then obtain the confidence limits

$$b_{21} \pm t_p \frac{s_2 \sqrt{1-r^2}}{s_1 \sqrt{n-2}}$$

for the unknown parameter β_{21}. The hypothesis that β_{21} has some a priori given value can be tested as in the preceding cases, by introducing this value into 14.4.13, and then comparing the resulting value of t with the t_p value for $n-2$ degrees of freedom.

Ex. 4. — For the shot material of Table 12.1.2, the sample characteristics as given in 12.5 are: $n = 96$, $s_1 = 11.209$, $s_2 = 16.414$, $r = 0.086$, $b_{21} = 0.126$. Do the data agree with the hypothesis that the true value of the regression coefficient is $\beta_{21} = 0$?

Substituting $\beta_{21} = 0$ in 14.4.13 we obtain $t = 0.837$. This falls short of the 5% value of t for $n-2 = 94$ degrees of freedom, so that there is no significant deviation from the hypothetical value, and the hypothesis $\beta_{21} = 0$ can be accepted, If we choose a lower level than 5%, the same conclusion holds a fortiori.

5. *The correlation coefficient.* — For the correlation coefficient, the situation is more complicated than in the preceding case. The sampling distribution of the observed correlation coefficient r will, in fact, take a simple form only in the particular case when the correlation coefficient of the population has the value $\varrho = 0$. According to 13.4.6, the variable

(14.4.14) $$t = \sqrt{n-2} \, \frac{r}{\sqrt{1-r^2}}$$

will in this case have a STUDENT distribution with $n-2$ degrees of freedom.

Since we do not want to discuss here the complicated distribution of r when $\varrho \neq 0$, we cannot deduce confidence limits for ϱ, but have to content ourselves with showing how the hypothesis $\varrho = 0$, i.e., the hypothesis that the variables are really uncorrelated, can be tested. In order to do this, we have only to find t from 14.4.14, and then compare as before with the t_p value for $n-2$ degrees of freedom. — Since $b_{21} = r s_2/s_1$, it is easily seen that the t value obtained in this way is identical with the one obtained when testing the hypothesis $\beta_{21} = 0$ according to 14.4.13.

Ex. 5. — For the shot material considered in Ex. 4, we obtain from 14.4.14 the same value $t = 0.837$ as in the preceding case, when we were testing the hypothesis $\beta_{21} = 0$. Thus the data do not indicate any significant correlation between the azimuth and height deviations.

14.5. Confidence intervals, II. — The method used in the preceding section for the determination of confidence intervals and the testing of certain statistical hypotheses is essentially based on the assumption that we know the exact sampling distributions of the various estimates involved. When this is not the case, we shall in most cases have to fall back on approximate methods applicable to large samples, where we can count on at least approximately normal distributions of our estimates.

Let us first consider the simple case of estimating the mean m of a variable not assumed to be normally distributed. As usual suppose that a sample of n observed values $x_1, \ldots, x_n$ is available. We then know (cf. 13.3) that for large n the sample mean $\bar{x} = \frac{1}{n} \sum x_i$ is asymptotically normal $(m, \sigma/\sqrt{n})$, where as usual σ denotes the s.d. of the variable.

Suppose for a moment that the value of σ is known. If λ_p denotes the $p \%$ value of the normal distribution (cf. 7.2), we then have approximately for large n

(14.5.1) $$P\left(-\lambda_p < \sqrt{n} \, \frac{\bar{x} - m}{\sigma} < \lambda_p \right) = 1 - \frac{p}{100} .$$

By the same transformation as in the case of 14.4.4 we obtain the equivalent relation

$$P\left(\bar{x} - \lambda_p \frac{\sigma}{\sqrt{n}} < m < \bar{x} + \lambda_p \frac{\sigma}{\sqrt{n}}\right) = 1 - \frac{p}{100}.$$

On the $p\,\%$ level of significance we then, as in the case of 14.4.5, have the confidence limits $\bar{x} \pm \lambda_p \dfrac{\sigma}{\sqrt{n}}$ for the unknown parameter m.

Now in most cases the population characteristic σ will be unknown. However, by the final remarks of 13.2, the corresponding sample characteristic s will have a mean that for large n is approximately equal to σ, and a s.d. tending to zero as $n \to \infty$. Accordingly, when n is large, it will be legitimate to replace σ by s in the last relation, thus obtaining the approximate confidence limits $\bar{x} \pm \lambda_p \dfrac{s}{\sqrt{n}}$ for m.

The method now used in the case of the mean m is generally applicable to any characteristic having an estimate which is asymptotically normal. Let ζ denote some population characteristic, and suppose the corresponding sample characteristic z to be asymptotically normal, with the mean ζ and a certain s.d. $D(z)$. For large n we then have approximately

$$P\left(-\lambda_p < \frac{z - \zeta}{D(z)} < \lambda_p\right) = 1 - \frac{p}{100},$$

and the same transformation as before gives the equivalent relation

$$P\left(z - \lambda_p D(z) < \zeta < z + \lambda_p D(z)\right) = 1 - \frac{p}{100}.$$

If $D(z)$ is known, we should then have the confidence limits $z \pm \lambda_p D(z)$ for the unknown parameter ζ.

In most cases, however, $D(z)$ will be unknown, since its expression will contain some unknown population characteristic, as found above in the case of $D(\bar{x}) = \sigma/\sqrt{n}$. When n is large, the same argument as in that case will lead us to replace, in the expression of $D(z)$, each population characteristic by the corresponding sample characteristic. At the same time we retain in the expression of $D(z)$, which may consist of several terms (cf., e.g., 13.2.3), only the term which is dominating for large n. In this way the exact expression of $D(z)$ is replaced by another expression that will be likely to agree approximately with $D(z)$ for large n. We shall denote this approximately valid expression by $d(z)$, and call it the *standard error of z in large samples*.

For the unknown characteristic ζ we obtain in this way, on the $p\,\%$ level, the approximate confidence limits

$$z \pm \lambda_p d(z),$$

which only contain quantities that can be numerically computed from the sample values.

When a certain characteristic z and its standard error $d(z)$ have been computed from a set of sample values, it will be convenient to state the result of the computation as $z \pm d(z)$. This will then imply that, by multiplying the quantity $d(z)$ with the factor λ_p corresponding to any given level $p\%$, we shall obtain approximate confidence limits for the unknown value ζ. Thus, e.g., the result of a computation of the mean $\bar{x}$ and its standard error will be given as $\bar{x} \pm s/\sqrt{n}$.

In order to test any hypothetical value $\zeta = \zeta_0$, we can now proceed in the same way as in the preceding section. We first compute the quantity

$$(14.5.2) \qquad \lambda = \frac{z - \zeta_0}{d(z)}.$$

Working on the given level $p\%$, we then look up the $p\%$ value λ_p of the normal distribution from Table II. If $|\lambda| > \lambda_p$, the deviation from the hypothetical value ζ_0 is significant on the level $p\%$, and the hypothesis $\zeta = \zeta_0$ should be rejected, and vice versa.

In order to apply these rules in practice, it will be necessary to know the standard errors in large samples of some of the most usual sample characteristics. From the formulae given in 13.2 we obtain, e.g.,

$$(14.5.3) \qquad d(\bar{x}) = \frac{s}{\sqrt{n}}, \quad d(s^2) = \frac{\sqrt{m_4 - s^4}}{\sqrt{n}}, \quad d(s) = \frac{\sqrt{m_4 - s^4}}{2s\sqrt{n}}.$$

The expressions for the standard errors will often be quite complicated. A considerable simplification can, however, be made if we are entitled to assume that the distribution of the population is at any rate not very far from the normal form. Under this condition we may, e.g., replace m_4 by $3s^4$ in the above expressions for $d(s^2)$ and $d(s)$, since for the normal distribution we have (cf. 7.2) $m_4 = 3s^4$. Under the same condition we have further, in respect of the measures of skewness and excess g_1 and g_2, and the correlation coefficient r,

$$(14.5.4) \qquad d(g_1) = \sqrt{\frac{6}{n}}, \quad d(g_2) = 2\sqrt{\frac{6}{n}}, \quad d(r) = \frac{1 - r^2}{\sqrt{n}}.$$

For an essentially non-negative variable, the ratio $V = s/\bar{x}$ is sometimes called the *coefficient of variation* of the sample. Under the same condition of approximate normality of the population we have approximately $E(V) = \sigma/m$, and

209

$$d\left(V\right) = \frac{V\sqrt{1+2\,V^2}}{\sqrt{2\,n}}.$$

Finally, for the frequency ratio f/n of a certain event in a series of n independent observations we find according to 6.2

$$d\left(\frac{f}{n}\right) = \sqrt{\frac{f\left(n-f\right)}{n^3}}.$$

It should be strongly emphasized that all the rules and formulae given in the present section are only approximately valid, and even this only for large samples. About the meaning of the word "large" in this connection, the remarks made in 13.3 will still hold. The formula for $d(f/n)$ can as a rule be safely used when f and $n-f$ both amount to at least 20.

Ex. 1. — Suppose that each individual in a large population, if asked a certain question, would give a definite answer "yes" or "no". For a public opinion poll, we draw a random sample of n individuals, among which we find f "yes". It is assumed that n is a large number, though small compared with the total number of individuals in the population. For the unknown proportion of "yes" in the total population, we then have the approximate confidence limits $\frac{f}{n} \pm \lambda_p \sqrt{\frac{f\left(n-f\right)}{n^3}}$. Any hypothetical value π_0 of the unknown proportion may now be tested as shown above.

Further, let f_1/n_1 and f_2/n_2 denote the frequency ratios of "yes" in samples drawn from two different populations. Do these data imply that the two populations have different proportions of "yes"?

Let us consider the hypothesis that the two unknown proportions are identical. If this hypothesis is true, and the common value of the two proportions is denoted by π, the frequency ratios f_1/n_1 and f_2/n_2 will both have the mean value π, and thus the difference $\Delta = f_1/n_1 - f_2/n_2$ will have the mean zero. Assuming the two samples to be independent, and making the same assumptions as before with respect to the magnitude of n_1 and n_2, the s.d. of Δ will be approximately $\sqrt{\frac{\pi\left(1-\pi\right)}{n_1} + \frac{\pi\left(1-\pi\right)}{n_2}}$. If the unknown proportion π is here replaced by the overall frequency ratio of "yes" in our data, which is f/n, where $f = f_1 + f_2$ and $n = n_1 + n_2$, we obtain for the standard error of Δ in large samples the expression $d\left(\Delta\right) = \sqrt{\frac{f\left(n-f\right)}{n\,n_1\,n_2}}$. If

$$\frac{|\Delta|}{d\left(\Delta\right)} = \sqrt{\frac{n\,n_1\,n_2}{f\left(n-f\right)}} \left|\frac{f_1}{n_1} - \frac{f_2}{n_2}\right| > \lambda_p,$$

the observed value of the difference Δ differs significantly on the level p % from the hypothetical value zero. Our hypothesis should then be rejected, i.e., our data imply a significant difference between the proportions of "yes" in the two populations. In the opposite case we should conclude that our data are consistent with the hypothesis of equal proportions. — Examples of this type very often occur in various fields of application. Cf. also 15.2, ex. 5.

Ex. 2. — On two different occasions, measurements of the stature of Swedish military conscripts have given the following figures:

	1915–16	1924–25
n	80,084	89,377
$\bar{x}$, in cm	171.80 ± 0.022	172.58 ± 0.020
s, in cm	6.15 ± 0.015	6.04 ± 0.014

The difference between the mean values is $+0.78 \pm \sqrt{0.022^2 + 0.020^2} = +0.78 \pm 0.030$. Thus there is a highly significant increase of the mean between the two investigations. For the difference between the two unknown population means we have the confidence limits $0.78 \pm 0.030 \, \lambda_p$. Taking, e.g., $p = 0.1\,\%$, these limits will be 0.68 and 0.88. — On the other hand, the difference between the two standard deviations is -0.11 ± 0.021, which implies a strongly significant decrease of the population value of σ.

Ex. 3. — For the normal distribution, the measures of skewness and excess, γ_1 and γ_2, are both equal to zero (cf. 7.2). In order to find out whether a given set of sample values may be regarded as drawn from a normally distributed population, we might thus test whether the corresponding sample characteristics, g_1 and g_2, differ significantly from zero. If either one shows a significant deviation, the hypothesis of normality should be rejected.

For the $n = 250$ ash percentages of Table 12.1.1, we have in 12.5 found the values $g_1 = 0.062$ and $g_2 = 0.200$. From the histogram in Fig. 26 it is evident that the distribution is at any rate not very far from normal, so that we are entitled to use the expressions 14.5.4 for the standard errors. By means of these expressions we find $g_1 = 0.062 \pm 0.155$ and $g_2 = 0.200 \pm 0.310$. Thus neither value shows a significant deviation from zero. Cf. further 15.2, ex. 2.

14.6. General remarks on the testing of statistical hypotheses. — Some
preliminary remarks on this subject have already been made in 11.1 and 11.3. In the preceding sections of the present chapter, we have repeatedly encountered the problem of testing, on the basis of a set of sample values, some hypothesis of the type $\zeta = \zeta_0$, where ζ is an unknown population characteristic, while ζ_0 is a given number. We have shown in various instances how it is possible to construct tests of an hypothesis of this kind, such that the probability of rejecting the hypothesis in a case when it is in fact true has any given value $p\,\%$.

Now it is evident that this type of hypotheses is a very special one. In many applications, we shall be interested in testing hypotheses of various other kinds, such as the hypothesis that some unknown characteristic ζ takes any value belonging to some specified interval $\zeta_1 < \zeta < \zeta_2$, the hypothesis that the observed sample has been drawn from a normally distributed population, etc.

In respect of many of these more general hypotheses it will still be possible to find tests with the property that the probability of rejecting a true hypothesis has the a priori given value p %. If we apply such a test to a large number of cases where the hypothesis tested is in fact true, we shall thus in the long run take a wrong decision (i.e., by unjustly rejecting a true hypothesis) in about p % of all cases.

The trouble is, however, that in general there are for any given hypothesis a large number of different possible tests on the same level p %. We have already seen an example of this in 14.4, in respect of the test based on a confidence interval for the mean of a normal distribution, and we shall encounter a further example in connection with the discussion of the χ^2 test in the following chapter.

Given a hypothesis that we want to test, there will thus in general exist many different tests, all of which give the same probability p % of rejecting the hypothesis when it is true. How are we to choose between these tests?

In any case when we are concerned with testing a hypothesis, there are two possible decisions: we may decide to *accept* or to *reject* the hypothesis.[1]

In the former case we decide to assert that the hypothesis is true, while in the latter case we decide in favour of some of the alternative hypotheses that are considered a priori possible.

When, e.g., we are testing the hypothesis $m = m_0$ for a normal distribution, it is a priori assumed that the unknown distribution belongs to the family of all normal distributions. Accepting the hypothesis then means asserting that we are concerned with a distribution belonging to the sub-family consisting of all normal distributions with $m = m_0$ (whatever the value of σ), while rejecting means asserting that one of the a priori possible alternative hypotheses is true, i.e., that the distribution is normal, with $m \neq m_0$.

Now in both cases our decision may be wrong, since we may decide to reject a hypothesis which is in fact true, or to accept a false one. Consider now a family of different tests of the same hypothesis, all of which give the same probability p % of rejecting the hypothesis when it is true. It then seems reasonable that, choosing from among these tests, we should try to find one that gives *the smallest possible probability of accepting the hypothesis*

---

[1] As indicated in 11.3, there may also be the third possibility that we decide not to give any definite verdict until further observations have been made. The systematic development of this point of view leads to an important method mainly due to A. WALD and known as the *sequential* testing of hypotheses. For an account of this method, the reader is referred to WALD's *Sequential Analysis*.

in a case when it is in fact false (i.e., when one of the a priori possible alternative hypotheses is true). By acting in this way we shall, in fact, do what we can in order to reduce our risk of making a wrong decision.

The principle thus stated is fundamental for the modern theory of testing statistical hypotheses developed by J. NEYMAN, E. S. PEARSON, and others. It would fall outside the scope of the present book to give an account of the methods and results of this theory, for which we must refer to more advanced treatises, such as *Mathematical Methods*. However, the tests of various hypotheses discussed above have been given with due regard for the principle just stated. In the following chapter we shall consider a further important test, where we shall have the opportunity of making at least some tentative use of the principle, even though a complete discussion of its application would carry us too far.

The concept of the risk of making a wrong decision, which has been briefly touched above, may be further elaborated and applied to even more general cases of statistical decision problems than the testing of hypotheses. A general theory covering all these cases has been worked out in a series of papers by A. WALD, and summed up in his book *Statistical Decision Functions*.

14.7. Problems. — 1. A sample is drawn from a population known to have the fr.f. $f(x - m)$, where $f(x)$ is the fr.f. of the CAUCHY distribution, while m is an unknown parameter. Is the sample mean $\bar{x}$ a consistent estimate of m?

2. Consider the following three estimates of the mean of a rectangular population: a) the sample mean $\bar{x}$, b) the sample median z_2 (you may assume that n is odd), c) the arithmetic mean of the two extreme sample values: $\dfrac{x_{max} + x_{min}}{2}$. Show that all three are unbiased estimates, and compare their efficiency.

3. Find a 99.9% confidence interval for the atomic weight of iodine from the data given in 15.2, ex. 3.

4. Among 200 men aged 75 years, 20 died before attaining the age 76. Use 14.5 to find approximate 95% confidence limits for the unknown rate of mortality q_{75} (cf. 3.7, ex. 15).

5. Find, by means of a large sample, a $100\text{-}p$ % confidence interval for the parameter λ of a POISSON distribution.

6. In order to compare two machines for the manufacture of cotton yarn, ten samples from each machine were tested for tensile strength, the following values being obtained (data supplied by Mr. G. BLOM):

Machine A: 380, 382, 362, 382, 364, 392, 360, 358, 400, 402.

Machine B: 398, 440, 396, 428, 420, 446, 420, 420, 410, 390.

Assuming that the distribution of tensile strength for each machine is normal, with the same s.d. for both machines, find a 95% confidence interval for the difference between the mean values.

7. In order to determine the percentage of dichlormetanol in pulp, a certain analytic method is used. The result of each determination by means of this method may be regarded as a normally distributed variable, with a mean m equal to the true percentage, and a s.d. σ independent of the value of m. For each of 13 batches of pulp, two independent determinations were made, the following values being obtained (data supplied by Mr. G. BLOM):

Batch	Percentage 1	Percentage 2
1	.039	.013
2	.051	.045
3	.073	.085
4	.112	.119
5	.204	.208
6	.191	.181
7	.234	.265
8	.220	.205
9	.267	.260
10	.330	.300
11	.354	.355
12	.423	.460
13	.531	.554

Find from these data a 98% confidence interval for σ. (Hint: consider the differences between the two values obtained for each batch).

8. A population consisting of a large number of individuals is known to be a mixture of two populations A and B, in the unknown proportions p and $q = 1 - p$ respectively. In A, all individuals have the value zero of a certain characteristic, while in B this characteristic has a POISSON distribution with an unknown parameter λ. A sample of n values is drawn from the mixed population. Find maximum likelihood estimates of p and λ. Cf. 15.3, ex. 4.

9. Let ζ denote the unknown median of a distribution known to belong to the continuous type, and to have a unique median. Let x_i be the i-th order statistic of a sample of n values from this distribution. Show that for any i, j we have

$$P(x_i > \zeta) = \sum_{\nu=0}^{i-1} \binom{n}{\nu} \left(\frac{1}{2}\right)^n, \quad \text{and } P(x_j < \zeta) = \sum_{\nu=j}^{n} \binom{n}{\nu} \left(\frac{1}{2}\right)^n.$$

Assuming $i < j$, deduce that $P(x_i < \zeta < x_j) = \sum_{\nu=i}^{j-1} \binom{n}{\nu} \left(\frac{1}{2}\right)^n$, so that (x_i, x_j) is a confidence interval for the unknown median, the confidence coefficient being $\sum_{\nu=i}^{j-1} \binom{n}{\nu} \left(\frac{1}{2}\right)^n$.

Note that we cannot obtain in this way a confidence interval for any a priori given value of the confidence coefficient, but only for those that may be expressed in the above form. (When n is so large that the sum of binomial terms may be satisfactorily approximated by a normal integral, any given confidence coefficient may, of course, be approximately represented in this form.) In practice it is usually convenient to take $j = n - i + 1$, so that the confidence limits are the i-th observations from the top and from the bottom of the sample.

In a sample of 17 students, the median time required for taking a certain academic degree was 20 months, while the corresponding quartile values were 16 and 26 months. Find a confidence interval for the median of the distribution, based on the two quartile values, and compute the corresponding confidence coefficient.

10. $x_1, \ldots, x_n$ are the order statistics of a large sample from a continuous distribution with the d.f. $F(x)$. Use the method of 14.5 to find, for any given x, a 100-p% confidence interval for $F(x)$. Under what conditions would you consider the approximation involved satisfactory?

CHAPTER 15

THE χ^2 TEST

15.1. The case of a completely specified hypothetical distribution. — If we are given a set of sample values $x_1, \ldots, x_n$ of a random variable X, it will often be required to test the hypothesis that X has a given d.f. $F(x)$.

According to 12.3, the distribution of the sample may be regarded as a statistical image of the distribution of the variable. If we draw a sum polygon or a histogram representing our data, we may then compare this with the distribution curve or the frequency curve corresponding to the hypothetical d.f. $F(x)$, and so obtain at least a qualitative idea of the agreement between the two distributions. Examples of such diagrams are given in Figs. 24–27. However, in order to give a precise verdict, it will obviously be necessary to introduce some quantitative measure of the degree of deviation from the hypothetical distribution shown by our data. If this measure exceeds some appropriately fixed limit, we should then reject the hypothesis, and vice versa.

Such a deviation measure can be defined in various ways. We shall here only be able to consider one of these, viz., the measure leading to the important χ^2 test first introduced by K. PEARSON.

Let us first divide the total range of the variable X into a finite number, say r, of mutually exclusive groups. Let p_i denote the probability, according to the hypothetical d.f. $F(x)$, that the variable takes a value belonging to the i-th group, while f_i denotes the number of sample values falling in the same group. We then have

$$p_1 + p_2 + \ldots + p_r = 1,$$

(15.1.1)

$$f_1 + f_2 + \ldots + f_r = n.$$

If the hypothesis to be tested is true, f_i represents the frequency, in a series of n observations, of an event having the probability p_i, viz., the probability

215

that the variable takes a value belonging to the i-th group. Hence f_i has a binomial distribution (cf. 6.2) with the mean np_i, and it follows from 6.4 that when n is large f_i will be asymptotically normal $(np_i, \sqrt{np_i(1-p_i)})$.

Assuming that the hypothesis is true, and that n is large, we may thus expect the two sequences of numbers $f_1, \ldots, f_r$, and $np_1, \ldots, np_r$, to agree fairly well. As a comprehensive measure of the deviation of the *observed* frequencies f_i from the *expected* frequencies np_i, we now introduce the expression

$$(15.1.2) \qquad \chi^2 = \sum_1^r \frac{(f_i - np_i)^2}{np_i} = \sum_1^r \frac{f_i^2}{np_i} - n.$$

Writing

$$z_i = \frac{f_i - np_i}{\sqrt{np_i}},$$

z_i will be asymptotically normal, with a mean equal to zero, and we have

$$\chi^2 = \sum_1^r z_i^2,$$

where according to 15.1.1 the variables z_i satisfy the identity

$$(15.1.3) \qquad \sum_1^r z_i \sqrt{p_i} = 0.$$

Thus χ^2 is the sum of r asymptotically normal variables, which are subject to the constraint that they must satisfy the linear identity 15.1.3. Our first object is now to find the asymptotic distribution of χ^2. In 13.4 we have been concerned with a somewhat similar problem in connection with the distribution of the variable $\dfrac{ns^2}{\sigma^2} = \sum_1^n \left(\dfrac{x_i - \bar{x}}{\sigma}\right)^2$. This is the sum of the squares of n normally distributed variables, which are subject to one linear constraint, and we found that the effect of the constraint was to reduce the number of degrees of freedom in the resulting χ^2 distribution from n to $n-1$.

In the present case we have the sum of the squares of the r variables z_i, which are only asymptotically normal. Accordingly, we can only prove that the sum is *asymptotically* distributed in a χ^2 distribution, but the effect of the linear constraint proves to be the same as in the previous case. It is, in fact, possible to prove the following theorem:

If the hypothesis tested is true, the variable χ^2 defined by 15.1.2 has a distribution which, as n tends to infinity, tends to a χ^2 distribution with $r-1$ degrees of freedom.

216

For a complete proof of this theorem we must refer to *Mathematical Methods*, Chapter 30. — The simple result contained in the theorem will to a certain extent justify the somewhat arbitrary form of the measure χ^2 as defined by 15.1.2. It would, of course, have been quite as natural to consider any other expression of the form

$$\sum_1^r c_i^2 \, (f_i - n \, p_i)^2$$

with constant coefficients c_i, and the particular choice of coefficients in 15.1.2 has no other justification than the possibility to obtain in this way a simple asymptotic distribution.

As in the preceding chapter, we now choose a definite level of significance $p \%$ for the test that we are going to work out. Let χ_p^2 denote the $p \%$ value of the χ^2 distribution (cf. 8.1) for $r - 1$ degrees of freedom. If the hypothesis tested is true, we shall then have approximately for large values of n

$$(15.1.4) \qquad\qquad P\,(\chi^2 > \chi_p^2) = \frac{p}{100}\,.$$

When the numerical value of χ^2 for a given sample has been computed from 15.1.2, the hypothesis that the corresponding random variable has the given d.f. $F(x)$ can now be tested according to the following rule:

Whenever $\chi^2 > \chi_p^2$, we shall say that the sample shows a significant deviation from the hypothetical distribution, and the hypothesis will be rejected. On the other hand, when $\chi^2 \leq \chi_p^2$, the hypothesis will be accepted.

By applying this rule, we shall according to 15.1.4 have a probability of approximately $p \%$ of rejecting the hypothesis in a case when it is in fact true. Our test is thus for large n approximately on the $p \%$ level.

However, it should be noted that we test here, properly speaking, not the d.f. $F(x)$, but rather the probabilities p_i, or the group frequencies $n \, p_i$. Any d.f. giving the same values of the p_i will lead to the same value of χ^2, and will thus be accepted or rejected at the same time as F. When we assert that the hypothesis is true, what we really do assert is that the unknown d.f. of the variable is *some* d.f. giving the same p_i as our hypothetical F, and vice versa.

Let us now look at the frequency curve of the χ^2 distribution shown in Fig. 10, p. 80. The test just introduced implies that we shall reject the hypothesis whenever χ^2 takes a value falling in the right hand tail of the curve, $\chi^2 > \chi_p^2$. When the hypothesis tested is true, the probability of this event (disregarding the error of approximation) is $p \%$, which means that

217

we have the probability $p \%$ of rejecting the hypothesis when it is true, as already pointed out above. However, it is intuitive from the diagram that we can find infinitely many other intervals containing the same amount $p\%$ of probability mass in the χ^2 distribution. Let I denote any such interval, say the interval $a < \chi^2 < b$, so that we have for $r - 1$ degrees of freedom $P(a < \chi^2 < b) = p/100$. Then let us modify the above rule so that we reject the hypothesis whenever χ^2 takes a value belonging to I, and otherwise accept. Clearly we shall in this way obtain another test on the same level $p\%$. Thus each choice of an interval I yields a test on the desired level. We may, e.g., choose for I the *left* hand tail of the distribution, rejecting the hypothesis whenever $\chi^2 < \chi^2_{100-p}$, and otherwise accepting. This would still give us a test on the same level $p \%$, but one according to which the hypothesis would be rejected precisely in those cases where the deviation of the observed frequencies from the expected ones is particularly *small*, and accepted when the deviation is *large*, which certainly does not seem very reasonable. However, we shall obviously need some objective principle for choosing between all these possible tests on the same level.

Such a principle is, in fact, contained in the general remarks concerning the testing of hypotheses that have been made in 14.6. From among a family of tests, all giving the same probability $p \%$ of rejecting the hypothesis when it is true, we should try to choose one giving the smallest possible probability of accepting the hypothesis when it is false or, in other words, the largest possible probability of rejecting the hypothesis when it is false.

In the present case, our hypothesis implies that all the group frequencies f_i have the expected values np_i calculated from the hypothetical d.f. $F(x)$. If this hypothesis is false, some of the f_i will have mean values different from the corresponding np_i. It is readily seen that for any such f_i the mean value of the quantity $\dfrac{(f_i - np_i)^2}{np_i}$ will be of the same order of magnitude as n, and will thus be large for large values of n. *Thus the fact that the hypothesis is false will, generally speaking, tend to increase the value of χ^2.* The probability of having a χ^2 value falling in the right hand tail $\chi^2 > \chi^2_p$ will thus be larger when the hypothesis is false than when it is true, and this will at least make it plausible that, in order to maximize the probability of rejecting the hypothesis when it is false, we should choose to reject the hypothesis precisely when $\chi^2 > \chi^2_p$, and otherwise accept.

The χ^2 test, as deduced above, is based on the fact that each f_i is approximately normally distributed. When the test is applied in practice, and all expected frequencies np_i are $\geqq 10$, the approximation may be regarded as sufficiently good. When some of the np_i are < 10, it is usually advisable to pool the smaller groups, so that every group will contain at least 10 expected observations before the test is applied. When the observations are so few that this cannot be done, the results obtained by means of the χ^2 test must be used with caution.

When the number of degrees of freedom does not exceed 30, the χ_p^2 can be found from Table III. When there are $m > 30$ degrees of freedom, χ_p^2 can be computed from the approximate formula.

$$(15.1.5) \qquad \chi_p^2 = \tfrac{1}{2}\,(\sqrt{2\,m-1} + \lambda_{2\,p})^2$$

where $\lambda_{2\,p}$ is the $2\,p$ % value for the normal distribution, which is obtained from Table II. This formula is based on the fact that $\sqrt{2\,\chi^2}$ is approximately normal $(\sqrt{2\,m-1},\,1)$.

Ex. 1. — In one of MENDEL's classical experiments with peas, 556 peas were classified according to shape (round or angular) and colour (yellow or green), and the observed frequencies f_i given in the following table were obtained. Assuming that, in the language of the Mendelian theory of heredity, the properties "round" and "yellow" correspond to dominant genes, while "angular" and "green" are recessive, the combination of two dominant properties will have the probability $\dfrac{1}{16}$, while each of the two classes with one dominant and one recessive property has $\dfrac{3}{16}$, and the class with two recessive properties has $\dfrac{1}{16}$. This would give us the expected frequencies $n\,p_i$ shown in the table.

	Observed f_i	Expected $n\,p_i$	Difference $f_i - n\,p_i$
Round and yellow . .	315	312.75	2.25
Round and green . . .	108	104.25	3.75
Angular and yellow . .	101	104.25	− 3.25
Angular and green . .	32	34.75	− 2.75
Total	556	556.00	0.00

According to 15.1.2, these data give $\chi^2 = 0.470$. The number of classes is $r = 4$, so that we have to look up the χ^2 table for $r - 1 = 3$ degrees of freedom. The observed value of χ^2 lies between the 90 and 95% values, so that there is no significant deviation, and the hypothesis that the probabilities are in agreement with Mendelian theory can be accepted.

Ex. 2. — For $n = 2,880$ births the hour when the birth took place was observed, and the following data were obtained. The figures given for "hour" denote central values of one hour classes.

Hour:	0	1	2	3	4	5	6	7	8
Number:	127	139	143	138	134	115	129	113	126

Hour:	9	10	11	12	13	14	15	16	17
Number:	122	121	119	130	125	112	97	115	94

Hour:	18	19	20	21	22	23	Total
Number:	99	97	100	119	127	139	2,880

A glance at the figures will show that there seems to be a certain concentration of the cases during the night and early morning hours. Is this tendency significant? In other words, do the data show a significant deviation from the hypothesis that the hour of birth is uniformly distributed over the range from 0 to 24?

According to the hypothesis of a uniform distribution, each of the $r = 24$ classes has the probability $\frac{1}{24}$, so that the expected frequency for each class will be $2,880/24 = 120$. From 15.1.2 we then obtain $\chi^2 = 40.467$. For $r - 1 = 23$ degrees of freedom this value lies between the 1% and 2% values. According to the terminology introduced in 14.4, we should then say that our data show an "almost significant" deviation from the uniform distribution. Working, e.g., on the 1% level, the hypothesis of a uniform distribution can be accepted, while on the 5% level it should be rejected.

15.2. The case when certain parameters are estimated from the sample. — The case when the hypothetical d.f. $F(x)$ is completely specified is rather exceptional in the applications. More often we encounter cases where it is required to test the agreement between a set of sample values and some hypothetical distribution, the expression of which contains a certain number of unknown parameters, about the values of which we only possess such information as may be derived from the sample itself. We may, e.g., want to test the hypothesis that the variable with which we are dealing has *some* normal distribution, without specifying the values of the parameters m and σ.

In such a case, the hypothetical distribution is first *fitted to the observations*, i.e., the unknown parameters are estimated from the sample, and the estimates thus obtained are introduced into the expression of the distribution. In this way a completely specified d.f. is obtained, and we may then compute the deviation measure χ^2 from 15.1.2 as in the previous case.

However, it is necessary to observe that the significance of the value of χ^2 obtained in this way cannot be judged by the same rule as in the case treated in the preceding section. In fact, the probabilities p_i in the expression

$$\chi^2 = \sum_{1}^{r} \frac{(f_i - n\,p_i)^2}{n\,p_i}$$

are now no longer constants, but functions of the sample values, since they contain the estimated values of the parameters in the distribution. Say, in order to consider an extreme case, that we have r unknown parameters. We can then in general find values of these parameters satisfying the r equations $f_i - n p_i = 0$ $(i = 1, 2, \ldots, r)$, and if we use these values as our estimates we shall have $\chi^2 = 0$, so that in this case it is obvious that the theorem on the limiting distribution of χ^2 given in the preceding section is no longer true.

However, it is possible to show that under certain conditions a simple modification is enough to ensure the validity of the theorem even in the present case. Let us suppose that the hypothetical distribution contains s unknown parameters, where $s < r$, and that these parameters are estimated by the maximum likelihood method (cf. 14.3). In the expression for χ^2, the probabilities p_i should thus be computed after introduction of the maximum likelihood estimates for the s parameters into the hypothetical distribution. Subject to certain general regularity conditions, it is then possible to prove the following theorem first stated by R. A. FISHER (for a proof of the theorem, see *Mathematical Methods*, Chapter 30):

The theorem on the limiting distribution of χ^2 given in the preceding section remains true, provided that the number of degrees of freedom in the limiting distribution is replaced by $r - s - 1$.

Thus each parameter estimated from the sample reduces the number of degrees of freedom by one unit.

According to this theorem, the χ^2 test can be applied exactly as in the preceding section, provided only that the correct number of degrees of freedom is used. When s parameters have been estimated from the sample, the resulting χ^2 should be compared with the χ_p^2 value for $r - s - 1$ degrees of freedom, and the hypothesis tested should be rejected whenever $\chi^2 > \chi_p^2$, and otherwise accepted.

In order that the theorem given above should be true, the s parameters should be estimated by applying the maximum likelihood method to the observed frequencies f_i and the corresponding probalities p_i for the frequency groups actually used for the calculation of χ^2. The s unknown parameters, which enter into the expressions of the p_i, will then be determined so as to maximize the likelihood function

$$L = p_1^{f_1} \ldots p_r^{f_r}.$$

In the examples given below, we shall use the slightly different procedure of applying the maximum likelihood method to the original observations in each sample, before the grouping has been performed. This is, of course, only an approximation, but in many cases occurring in practice we may expect the error involved to be fairly small.

Ex. 1. — For the POISSON distribution, we have found in 14.3, ex. 2, that the maximum likelihood estimate of the parameter λ is the sample mean $\bar{x}$. When applying the χ^2 test to the hypothesis that a given sample agrees with some POISSON distribution, we thus have to replace the parameter λ by $\bar{x}$ in the expression $\dfrac{\lambda^i}{i!} e^{-\lambda}$ for the probability that the variable takes the value i.

For the blood corpuscle data in Table 12.1.1, we found in 12.5 the value $\bar{x} = 11.91$. For the computation of χ^2, we pool all classes with $i \leqq 7$ into one group, and similarly all classes with $i \geqq 17$. The expected frequencies will then exceed 10 for all groups, with only one exception. The probability p_i corresponding to the pooled group with $i \leqq 7$ will then, of course, be $\sum\limits_{0}^{7} \dfrac{\lambda^i}{i!} e^{-\lambda}$; where we have to take $\lambda = 11.91$, and similarly for the group with $i \geqq 17$. The computation of χ^2 is shown in the following table.

Group	f_i	$n p_i$	$\dfrac{(f_i - n p_i)^2}{n p_i}$
0—7	17	15.8032	0.0906
8	13	11.4081	0.2221
9	14	15.0967	0.0797
10	15	17.9802	0.4940
11	15	19.4676	1.0253
12	21	19.3216	0.1458
13	18	17.7016	0.0050
14	17	15.0590	0.2502
15	16	11.9568	1.3672
16	9	8.9004	0.0011
$\geqq 17$	14	16.3048	0.3258
Total	169	169.0000	4.0068

The number of classes is here $r = 11$, and we have estimated $s = 1$ parameter from the sample, so that there are $r - s - 1 = 9$ degrees of freedom. The observed value $\chi^2 = 4.0068$ lies between the 90 and 95% values, so that the agreement with the POISSON distribution is very good.

Ex. 2. — For the normal distribution, the maximum likelihood method gives according to 14.3 the estimates $\bar{x}$ for the parameter m, and s for σ. If the limits of the class interval numbered i are a and b, the corresponding probability p_i becomes, after the introduction of the estimates for the parameters,

$$p_i = \frac{1}{s\sqrt{2\pi}} \int_a^b e^{-\frac{(t-\bar{x})^2}{2s^2}} \, dt = \frac{1}{\sqrt{2\pi}} \int_{\frac{a-\bar{x}}{s}}^{\frac{b-\bar{x}}{s}} e^{-\frac{\lambda^2}{2}} \, d\lambda = \Phi\left(\frac{b-\bar{x}}{s}\right) - \Phi\left(\frac{a-\bar{x}}{s}\right).$$

Thus after having computed the standardized class limits $\dfrac{a-\bar{x}}{s}$ and $\dfrac{b-\bar{x}}{s}$, the p_i can be found directly from the values of $\Phi(x)$ given in Table I.

For the ash percentage data of Table 12.1.1, we found in 12.5 the estimates $\bar{x} = 17.015$ and $s = 2.635$. In 14,5, ex. 3, we have seen that for this material neither g_1 nor g_2 shows any significant deviation from normality. In order to apply the χ^2 test to the hypothesis that the distribution is normal, we pool the four lowest classes in the table, and also the five highest. All expected frequencies will then exceed 10, and the computation of χ^2 will be as shown in the following table.

Standardized class limits	f_i	$n\,p_i$	$\dfrac{(f_i - n\,p_i)^2}{n\,p_i}$
> -1.524	16	16.012	0.000
$-1.524, \quad -1.144$	13	15.640	0.446
$-1.144, \quad -0.765$	27	24.023	0.369
$-0.765, \quad -0.385$	28	31.878	0.472
$-0.385, \quad -0.006$	39	36.850	0.125
$-0.006, \quad 0.374$	42	37.015	0.671
$0.374, \quad 0.753$	34	32.080	0.115
$0.753, \quad 1.133$	19	24.285	1.150
$1.133, \quad 1.512$	14	15.875	0.222
$> \quad 1.512$	18	16.342	0.168
Total	250	250.000	3.738

We have here $r = 10$ groups, and $s = 2$ estimated parameters, so that there are $r - s - 1 = 7$ degrees of freedom. The observed value $\chi^2 = 3.738$ lies between the 80 and 90% values, so that the material agrees well with the hypothesis of a normal distribution.

In this example the probabilities p_i have been computed by means of linear interpolation between the $\Phi(x)$ values given in Table I. The resulting value of χ^2 is, of course, affected with a certain error due to this rather crude method of computation. However, the exact value of χ^2 as computed by means of a more extensive table of $\Phi(x)$ is $\chi^2 = 3.713$ so that the error is really not considerable.

A comparison between the histogram for the data discussed in this example and the normal frequency curve corresponding to the estimated values of the parameters has been given in Fig. 26, p. 169.

Ex. 3. — In $n = 52$ experimental determinations of the atomic weight of iodine, the following values were obtained (data from J. L. COOLIDGE, *Mathematical Probability*, Oxford 1925):

$10^3 (x - 126.980)$	f	$10^3 (x - 126.980)$	f	$10^3 (x - 126.980)$	f
13	2	4	2	− 5	2
12	0	3	4	− 6	1
11	2	2	1	− 7	3
10	2	1	1	− 8	1
9	1	0	3	− 9	0
8	3	− 1	3	− 10	0
7	4	− 2	0	− 11	2
6	3	− 3	4	− 12	0
5	5	− 4	2	− 13	1
				Total	52

From these data we obtain, using the expressions 14.5.3 and 14.5.4 for the standard errors, $\bar{x} = 126.98177 \pm 0.00090$, $s = 0.00646 \pm 0.00063$, $g_1 = -0.335 \pm 0.340$, and $g_2 = -0.714 \pm 0.680$. Thus neither g_1 nor g_2 shows any significant deviation from the normal value zero.

If we want to apply the χ^2 test to this example, we shall not be able to have more than four groups, in order to have a sufficient number of expected observations in each group. We may, e.g., group the observations as shown in the following table.

Class limits	f_i	$n p_i$	$\dfrac{(f_i - n p_i)^2}{n p_i}$
<126.9755	10	8.648	.211
$126.9755 - 126.9815$	13	16.484	.736
$126.9815 - 126.9875$	19	17.092	.213
>126.9875	10	9.776	.005
Total	52	25.000	1.165

The $n p_i$ are computed in the same way as in the preceding example, and we obtain $\chi^2 = 1.165$. Since there is $4 - 2 - 1 = 1$ degree of freedom, it follows from Table III that the material does not show any significant deviation from normality.

However, when the number of groups is so small as in this case, it is necessary to remember that, as pointed out above, the χ^2 test is in reality not more than a test of the group frequencies. In the present case we have found that there is no significant deviation from the expected frequencies computed from the normal distribution, but since there are only four groups, this cannot be regarded as a very strong evidence of a strictly normal form of the distribution.

A graphical comparison (on probability paper) between the sum polygon of the observations and the normal distribution curve corresponding to the estimated values of the parameters has been given in Fig. 25, p. 168.

Ex. 4. — The χ^2 test can also be used for testing the hypothesis that a number of different samples correspond to variables having *the same distribution*, without specifying the distribution. As an example we shall consider the case when k series of $n_1, \ldots, n_k$

observations respectively have been made, and the observed frequencies, $f_1 \ldots, f_k$ have been obtained for a certain event A. Are these data consistent with the hypothesis that A has a constant, though unknown probability p throughout all the observations?

The total number of observations is $n = \Sigma n_i$, and the total frequency of A is $f = \Sigma f_i$. Assuming that the hypothesis of a constant probability p is true, the maximum likelihood method gives (cf. 14.3, ex. 1) the estimate $p^* = f/n$.

For the calculation of the deviation measure χ^2, each of the k series of observations gives two groups, containing respectively the cases where A has occurred, and those where A has not occurred. The corresponding expected frequencies in the i-th series are $n_i p$ and $n_i(1 - p)$. Replacing here by p the estimate p^*, we find that the contribution to χ^2 from this series will be

$$\frac{(f_i - n_i p^*)^2}{n_i p^*} + \frac{(n_i - f_i - n_i q^*)^2}{n_i q^*} = \frac{(f_i - n_i p^*)^2}{n_i p^* q^*},$$

where $q^* = 1 - p^*$. Summing for all the k series we obtain

$$\chi^2 = \sum_1^k \frac{(f_i - n_i p^*)^2}{n_i p^* q^*} = \frac{1}{p^* q^*} \sum_1^k \frac{f_i^2}{n_i} - n \frac{p^*}{q^*}.$$

In each series we have $r = 2$ groups, and thus $r - 1 = 1$ degree of freedom. All the k series together, where $s = 1$ parameter has been estimated from the data, thus give us $k - 1$ degrees of freedom. The resulting χ^2 value will accordingly have to be judged by comparison with the values in Table III for $k - 1$ degrees of freedom.

The following table gives the number of new-born boys and girls in Sweden for each of the $k = 12$ months of the year 1935.

	Month												Total
	1	2	3	4	5	6	7	8	9	10	11	12	
Boys	3,743	3,550	4,017	4,173	4,117	3,944	3,964	3,797	3,712	3,512	3,392	3,761	45,682
Girls	3,537	3,407	3,866	3,711	3,775	3,665	3,621	3,596	3,491	3,391	3,160	3,371	42,591
Total	7,280	6,957	7,883	7,884	7,892	7,609	7,585	7,393	7,203	6,903	6,552	7,132	88,273

On the hypothesis that there is during all twelve months a constant probability that a new-born child will be a boy, the estimated value of this probability will be $p^* = 45,682/$ $/88,273 = 0.5175082$. From the above formula we then find $\chi^2 = 14.986$ with $k - 1 = 11$ degrees of freedom, and it will be seen from Table III that the agreement with the hypothesis of a constant probability is quite satisfactory.

Ex. 5. — In the particular case when $k = 2$ in the preceding example, we have a *four-fold table* of the following appearance.

f_1	f_2	f
$n_1 - f_1$	$n_2 - f_2$	$n - f$
n_1	n_2	n

In this case the above expression for χ^2 is readily brought on the form

$$\chi^2 = \frac{n\,n_1\,n_2}{f\,(n-f)}\left(\frac{f_1}{n_1} - \frac{f_2}{n_2}\right)^2,$$

where there is only $k - 1 = 1$ degree of freedom. The hypothesis to be tested asserts that the two frequency ratios f_1/n_1 and f_2/n_2 correspond to the same unknown probability p. It is interesting to compare the test given here with the test of the same hypothesis given in 14.5, ex. 1. Since the χ^2 distribution with one degree of freedom is (cf. 8.1) identical with the distribution of the square of a variable which is normal $(0,1)$, the two tests are clearly identical.

Among $n_1 = 979$ persons in the age group $25 - 30$ years, who during a certain calendar year had signed life insurance with normal premiums, there occurred $f_1 = 48$ deaths during the ten first years of insurance. Among $n_2 = 140$ persons in the same age group, who during the same year had signed insurance with increased premiums on account of impaired health, there occurred during the same period $f_2 = 13$ deaths. The two frequency ratios for deaths are $f_1/n_1 = 48/979 = 0.049$ and $f_2/n_2 = 13/140 = 0.093$. Are they significantly different?

For the fourfold table corresponding to this example, the above formula gives $\chi^2 = 4.565$. According to Table III, this is slightly above the 5% value for one degree of freedom, so that the difference is "almost significant". Working on the 5% level, we should accordingly reject the hypothesis that the two groups have the same probability of dying during the period considered.

15.3. Problems. — 1. 209 offspring of a cross experiment with Drosophila melanogaster were classified according to the following table (from BONNIER-TEDIN, *Biologisk variationsanalys*):

Bristles	Eyes	
	Normal	Reduced
Normal.	122	41
Reduced	40	6

According to genetic theory, the probabilities of the four classes should be

$$\frac{9}{16}, \quad \frac{3}{16},$$

$$\frac{3}{16}, \quad \frac{1}{16}.$$

Test the agreement of the data with this theoretical model.

2. From a human population, 1000 individuals were selected at random and classified according to the following table (from MOOD, *Theory of Statistics*):

	Male	Female
Normal.	442	514
Colour-blind . .	38	6

According to a genetic model, the probabilities of the four classes should be

$$\tfrac{1}{2}p, \quad \tfrac{1}{2}p^2 + pq,$$

$$\tfrac{1}{2}q, \quad \tfrac{1}{2}q^2,$$

where $q = 1 - p$ is the unknown proportion of defective genes in the population. Estimate q by the maximum likelihood method, and test the agreement of the data with the model.

3. The distribution of mean temperatures (centigrades) in Stockholm for the month of June during the 100 years 1841–1940 is shown below (cf. also Fig. 24, p. 167):

Temperature	Frequency
− 12.4	10
12.5 − 12.9	12
13.0 − 13.4	9
13.5 − 13.9	10
14.0 − 14.4	19
14.5 − 14.9	10
15.0 − 15.4	9
15.5 − 15.9	6
16.0 − 16.4	7
16.5 −	8
Total	100

Could these data be regarded as a sample from a normal distribution?

4. 4,075 widows receiving pensions from a certain pension fund were classified according to their number of children entitled to support from the fund. The following figures were obtained (from the Annual Report for 1952 of the pension fund S.P.P., Stockholm):

Number of children	Number of widows
0	3,062
1	587
2	284
3	103
4	33
5	4
6	2
Total	4,075

Fit the distribution considered in 14.7, ex. 8, to these data, find the maximum likelihood estimates of the parameters, and test the agreement by the χ^2 test.

5. The numbers π and e have been determined to 2000 decimal places (cf. Mathematical Tables and other Aids to Computation, 1950, p. 11). Each of the two numbers thus gives 10 groups of 200 digits. Assuming that each group of 200 digits may be regarded as a random sample from an infinite population containing the digits $0, 1, \ldots, 9$ in equal proportions, we may compute a value of χ^2 for each group according to the formula

$$\chi^2 = \frac{1}{20} \sum_0^9 (f_i - 20)^2,$$

where f_i is the frequency of the digit i in the group. In this way the following χ^2 values were obtained.

Group	π	e
1	6.8	8.1
2	7.4	12.9
3	6.8	12.8
4	10.1	10.0
5	7.5	5.5
6	7.1	2.0
7	4.4	10.8
8	6.2	8.2
9	8.4	4.8
10	5.1	7.8

If our hypothesis is true, the 20 observed values of χ^2 would constitute a random sample from a population having the χ^2 distribution with 9 degrees of freedom. By Table III, the 50% value, i.e., the median of this distribution is 8.343. In the above table, we find 14 values below the median, and 6 values above, as against 10 expected values on each side of the median. Is the agreement satisfactory? — The sum of all twenty values, 152.7, may be regarded as a single observed value of χ^2 with $20 \cdot 9 = 180$ degrees of freedom. Use 15.1.5 to judge the significance of this value.

CHAPTER 16

SOME FURTHER APPLICATIONS

16.1. Theory of errors. I. Direct observations. — Any method for the numerical determination of physical magnitudes is more or less liable to error. If repeated observations are made on the same physical magnitude, we shall practically always find that the value obtained will vary from observation to observation, and these fluctuations are usually ascribed to the presence of *errors of observation*. The object of the theory of errors, which was founded by GAUSS and LAPLACE, is to work out methods for estimating the numerical values of the required magnitudes by means of a given set of observations, and also to make it possible for the observer to arrive at some conclusion with respect to the degree of precision of the estimates obtained.

Let us assume that a certain method of observation or measurement is given, and that we may feel justified in disregarding from the beginning any error which might be due to incorrect use of the method, such as incorrect manipulation of instruments, errors of numerical computation, and the like. The remaining errors may be roughly classified in two large groups, known respectively as *systematic* and *random errors*. Even though it seems difficult to make this classification entirely precise, it will tend to clarify the discussion of the nature of the errors, and we propose to say a few words about each of the two groups.

A *systematic error* may be described as an intrinsic error of the method of observation itself. An error of this kind will thus always be present, and have a constant magnitude, at any application of the same method. Say, e.g., that repeated determinations of the weight of a given piece of metal are performed by means of a set of weights, all of which are too light by a constant percentage. This bias of the set of weights will then give rise to a systematic error in our measurements, each observed value of the weight of the piece of metal being affected with an error proportional to the bias of the set of weights. Errors of this kind will have to be eliminated by a careful analysis of the methods of observation used. Obviously they will not lend themselves to probabilistic treatment, and accordingly we shall assume in the sequel that we are dealing with observations which are free from systematic errors.

Any method of observation will be more or less affected with *random errors*. These will be due to a large number of irregular and fluctuating causes that cannot be completely controlled by the observer, such as vibrations of the instrument used for measurements, errors in reading a scale, uncontrollable variations of temperature, currents of air, etc. The errors due to causes of such kinds will be irregularly fluctuating from observation to observation, and their values cannot be accurately predicted. Accordingly errors of this type may be regarded as random variables, and treated by the methods of probability theory.

In any method of observation, a large number of potential causes of random errors will be present. If we may assume that each of these causes produces a certain *elementary error*, and that the various elementary errors are independent random variables, which add together to make up the resulting total error,

229

the central limit theorem (cf. 7.4 and 7.7) will show that the total error is approximately normally distributed. Now, these assumptions clearly represent a certain oversimplification of the facts, but on the other hand the central limit theorem is known to hold even under more general conditions, and in fact statistical experience shows, as we have several times pointed out (cf. 7.7, 13.4, and 15.2, ex. 3), that sets of repeated experimental measurements will often be approximately normally distributed.

Accordingly the theory of errors is usually built on the simplifying assumption that errors of measurement or observation can be regarded as normally distributed random variables. The unknown *true value* of an observed magnitude is then identified with the mean of the corresponding normal distribution, while the standard deviation of the distribution will characterize the degree of precision in the measurements. — It should be pointed out, however, that some of the results of the theory as given below are, in reality, independent of the assumption of the normal distribution.

Consider first the simple case when we are given a set of n measurements $x_1, \ldots, x_n$ on some observed magnitude X, and it is required to use these measurements to form an estimate of the unknown true value m of X, and to judge the degree of precision of the estimate obtained. Obviously this coincides with the problem of estimating the mean m of a normal distribution from a set of sample values, which has been treated in 14.4 and 14.5 by two different methods. Introducing as usual the notations

$$\bar{x} = \frac{1}{n} \sum_1^n x_i, \quad s^2 = \frac{1}{n} \sum_1^n (x_i - \bar{x})^2,$$

the exact method of 14.4 gives us, on the level of significance p percent, the confidence limits

(16.1.1)
$$\bar{x} \pm t_p \frac{s}{\sqrt{n-1}}$$

for the unknown true value m. Here the quantity t_p will have to be taken from Table IV, with $n-1$ degrees of freedom. According to the approximate method of 14.5, the same confidence limits will, for large n, be approximately given by the expression

(16.1.2)
$$\bar{x} \pm \lambda_p \frac{s}{\sqrt{n}},$$

where λ_p has to be taken from Table II. This approximate method may be used (with due care) even when the assumption of the normal distribution of errors seems doubtful.

By 13.2.1, the standard deviation of $\bar{x}$ is $\sqrt{\sigma^2/n}$, and if we estimate the unknown variance σ^2 by means of the corrected sample variance $\dfrac{n}{n-1}s^2$ (cf. 13.2.2), we obtain for the s.d. of $\bar{x}$ the estimate $s/\sqrt{n-1}$, which appears as coefficient of t_p in 16.1.1. If n is large, this will be approximately equal to $s/\sqrt{n}$, the coefficient of λ_p in 16.1.2. The latter expression for the standard error of $\bar{x}$ in large samples has been given above in 14.5.3.

The result of the set of measurements may be given in the conventional form explained in 14.5, as $\bar{x} \pm s/\sqrt{n-1}$ or, for large n, as $\bar{x} \pm s/\sqrt{n}$. Thus the result of the set of $n = 52$ experimental determinations of the atomic weight of iodine discussed in 15.2, ex. 3., would be given as 126.9818 ± 0.00090, whether we use 16.1.1 or 16.1.2. The 95 % confidence limits will here be 126.9800 and 126.9836, whichever of the two expressions is used.

The problem of estimating m from a set of measurements has sometimes been treated as an application of BAYES' theorem (cf. 3.6). The unknown true value of the observed magnitude was then conceived as a random variable, and it was assumed that there exists an *a priori probability* $\pi(m)\, dm$ of the event that this variable should have a value between m and $m + dm$. By means of BAYES' theorem it was then possible to obtain an expression of the *a posteriori probability* of the same event, *after having observed the values* $x_1, \ldots, x_n$. By integration it was then possible to find the probability of the event that the true value was situated between any given limits m_0 and m_1.

The weakness of this method is, in the first place, due to the fact that usually the a priori frequency function $\pi(m)$ will be entirely unknown. Moreover, in many cases it seems extremely doubtful whether it is at all legitimate to regard m as a random variable. Usually it is more natural to regard m simply as a fixed, though unknown constant which appears as a parameter in the X distribution from which sample values are taken. In such a case, there do not seem to be any reasons for assuming that the value of m has been determined by means of some procedure resembling a random experiment. If this position is taken, there does not exist any a priori probability distribution of m, and BAYES' theorem cannot be used. On the other hand, the method of confidence intervals given above can be used whether m is a random variable or not.

Suppose that a certain number, say k, of independent sets of measurements have been made, one on each of k different magnitudes. The normal distributions involved will then in general have k different mean values $m_1, \ldots, m_k$. However, it sometimes occurs that there will be reasons for assuming that the s.d. has the same (unknown) value σ for all k distributions. This will, e.g., often occur when experimental determinations are made on different unknown magnitudes by means of the same experimental method. It is then possible to obtain, for each of the unknown

231

quantities m_i, a better estimate if all k sets of measurements are used than by only using the measurements in the set directly connected with m_i. Let the k sets include $n_1, \ldots, n_k$ measurements respectively, let the observed mean values and variances be $\bar{x}_1, \ldots, \bar{x}_k$ and $s_1^2, \ldots, s_k^2$, and write $N = \sum_1^k n_i$. According to 13.4 the variable $n_i s_i^2/\sigma^2$ can be represented as a sum of the squares of $n_i - 1$ independent variables, each of which is normal $(0, 1)$, and thus this variable has a χ^2 distribution with $n_i - 1$ degrees of freedom. Since the measurements in any two different sets are assumed to be mutually independent, it follows that the variable $\frac{1}{\sigma^2} \sum_1^k n_i s_i^2$ can be represented as a sum of $\sum_1^k (n_i - 1) = N - k$ variables with the same properties as before. The degrees of freedom will then add together (cf. 13.4), so that the variable $\frac{1}{\sigma^2} \sum_1^k n_i s_i^2$ will have a χ^2 distribution with $N - k$ degrees of freedom. In the same way as in connection with 13.4.3 we then find that for any $j = 1, 2, \ldots, k$ the variable

$$ t = \sqrt{n_j(N-k)} \, \frac{\bar{x}_j - m_j}{\sqrt{\sum_1^k n_i s_i^2}} $$

will have a STUDENT distribution with $N - k$ degress of freedom. For the unknown true value m_j, we then obtain as in 14.4 the confidence limits

$$ (16.1.3) \qquad \bar{x}_j \pm t_p \sqrt{\frac{\sum_1^k n_i s_i^2}{n_j(N-k)}}, $$

where t_p should be taken from Table IV, with $N - k$ degrees of freedom.

Usually the formula 16.1.3 will give a better estimate of m_j, i.e., a shorter confidence interval for a given value of p, then would be obtained from the formula 16.1.1, using only the measurements in the j-th set, which is directly connected with m_j. The result of the j-th set of measurements should in this case be given in the same conventional form as before, as

$$ \bar{x}_j \pm \sqrt{\frac{\sum_1^k n_i s_i^2}{n_j(N-k)}}. $$

In practical applications, the case when $n_1 = \ldots = n_k = 2$ often occurs. In this case, one pair of measurements has been made on each of k different magnitudes, and all $N = 2k$

measurements are assumed to have the same standard deviation. If the observed values in the j-th pair are denoted by x_{j1} and x_{j2}, we shall have

$$\bar{x}_j = \tfrac{1}{2}(x_{j1} + x_{j2}) \ \text{ and } \ s_j^2 = \tfrac{1}{4}(x_{j1} - x_{j2})^2.$$

The confidence limits for m_j will be

$$\bar{x}_j \pm \tfrac{1}{2} t_p \sqrt{\frac{1}{k} \sum_1^k (x_{i\,1} - x_{i\,2})^2},$$

where t_p has $2k - k = k$ degrees of freedom. If, in this case, confidence limits for m_j are determined from 16.1.1, using only the measurements in the j-th pair, we shall obtain the values $\bar{x}_j \pm \tfrac{1}{2} t_p |x_{j1} - x_{j2}|$ where, now, t_p will only have $2 - 1 = 1$ degree of freedom. The coefficients of t_p in both expressions for the confidence limits for m_j may be expected to be of the same order of magnitude. For a given value of p, the factor t_p will, on the other hand, decrease when the number of degrees of freedom increases. Thus for $p = 1\%$ Table IV gives $t_p = 63.7$ for one degree of freedom, but $t_p = 3.2$ for $k = 10$ degrees of freedom. As soon as the number of pairs of measurements is at least moderately large, we shall thus in general obtain a better estimate, i.e., a shorter confidence interval for m_j, by using all the available measurements instead of only the two measurements directly concerned with m_j.

We shall finally consider the following problem, which will appear, e.g., in geodetic applications. In a series of n measurements on a certain magnitude, the unknown true value of which is m, the values $x_1, \ldots, x_n$ have been found. The measurements have not all been performed according to the same method, and the corresponding standard deviations $\sigma_1, \ldots, \sigma_n$ are not known, but it is assumed that the ratio between any two of the σ_i is known, so that we may write

$$\sigma_i^2 = \frac{\sigma^2}{p_i},$$

where σ^2 is an unknown constant, while $p_1, \ldots, p_n$ are known positive quantities. For reasons that will appear below, p_i will be denoted as the *weight* of the i-th observation. It is now required to estimate the true value m.

In this case $x_1, \ldots, x_n$ are independent normal variables with the same mean m, the variance of x_i being σ^2/p_i. In order to estimate m according to the maximum likelihood method (cf. 14.3), we have to find m so as to maximize the likelihood function

$$L(x_1, \ldots, x_n ; m, \sigma) = \frac{\sqrt{p_1 \ldots p_n}}{\sigma^n (2\pi)^{n/2}} e^{-\frac{1}{2\sigma^2} \sum_1^n p_i (x_i - m)^2}.$$

233

By a simple calculation, we obtain as the solution of this problem the *weighted arithmetic mean*

$$\bar{\bar{x}} = \frac{p_1 x_1 + \cdots + p_n x_n}{p_1 + \cdots + p_n}.$$

$\bar{\bar{x}}$ is a linear function of the random variables $x_1, \ldots, x_n$, and we easily find

$$E(\bar{\bar{x}}) = m, \quad D^2(\bar{\bar{x}}) = \frac{\sigma^2}{\sum_1^n p_i}.$$

The form of the expression found for the estimate $\bar{\bar{x}}$ will justify the use of the name "weights" for the constants p_i. It will be seen from the expression for $D^2(\bar{\bar{x}})$ that the weight of the mean value $\bar{\bar{x}}$ is the sum of the weights of all the measurements x_i. In the particular case when all the weights are equal, $\bar{\bar{x}}$ will obviously coincide with the ordinary arithmetic mean $\bar{x}$.

In order to find confidence limits for m we form the expression

$$\bar{s}^2 = \frac{\sum_1^n p_i (x_i - \bar{\bar{x}})^2}{\sum_1^n p_i}.$$

By a simple modification of the argument used in 13.4 for the deduction of the joint distribution of $\bar{x}$ and s^2, we obtain here the following perfectly analogous result: $\bar{\bar{x}}$ and $\bar{s}^2$ are independent, and $\bar{\bar{x}}$ is normal $\left(m, \sigma \Big/ \sqrt{\sum_1^n p_i} \right)$, while $\bar{s}^2 \sum_1^n p_i / \sigma^2$ has a χ^2 distribution with $n-1$ degrees of freedom. Exactly in the same way as in connection with 13.4.3 we further find that the variable

$$t = \sqrt{n-1}\, \frac{\bar{\bar{x}} - m}{\bar{s}}$$

has a STUDENT distribution with $n-1$ degrees of freedom, and finally we obtain as in 14.4.5 the confidence limits

$$\bar{\bar{x}} \pm t_p \frac{\bar{s}}{\sqrt{n-1}}$$

for the unknown true value m. Here t_p has to be taken from Table IV, for the level p %, and $n-1$ degrees of freedom. The result of the series of measurements should be given in the same conventional form as above:

$$\bar{\bar{x}} \pm \frac{\bar{s}}{\sqrt{n-1}}.$$

16.2. Theory of errors. II. Indirect observations.

— In many cases, the magnitudes that we want to determine are not available for direct measurement. Sometimes it will, however, be possible to measure directly certain *functions* of these magnitudes, and it will then be required to use these measurements in order to estimate the unknown true values of the required magnitudes themselves. If our measurements are fairly precise, and if the functions concerned are sufficiently regular with respect to continuity and derivability, these functions may be regarded as approximately linear. We shall here restrict ourselves to the case when the functions measured are strictly linear, and the results obtained will then be at least approximately true even under more general circumstances.

Suppose that it is required to determine k magnitudes, the unknown true values of which are $m_1, \ldots, m_k$. We observe a certain number $n > k$ of linear functions of these magnitudes, the true values of which are

$$f_1 = a_{11} m_1 + a_{12} m_2 + \cdots + a_{1k} m_k,$$
$$f_2 = a_{21} m_1 + a_{22} m_2 + \cdots + a_{2k} m_k,$$

(16.2.1) $$\cdots \cdots \cdots \cdots \cdots$$

$$f_n = a_{n1} m_1 + a_{n2} m_2 + \cdots + a_{nk} m_k,$$

where the coefficients a_{ir} are assumed to be known constants. In order to exclude the possibility of a too close dependence between the functions f_i we shall assume that it is possible to find among these functions at least one group of k functions, such that the corresponding determinant of the coefficients a_{ir} is different from zero.

Suppose that we have performed one measurement on each f_i, and that the value x_i obtained at this measurement can be regarded as an observed value of a normally distributed random variable with the unknown mean f_i. As in the last problem of the preceding section, we shall assume that the s.d:s of our n measurements may be unequal, but that the corresponding weights p_i are known, so that we have

$$E(x_i) = f_i, \quad D^2(x_i) = \frac{\sigma^2}{p_i},$$

235

where σ^2 and the f_i are unknown, while the p_i are known, We also assume as before that all the n measurements are mutually independent.

In order to estimate the unknown true values m_i according to the maximum likelihood method, we form the likelihood function of the variables $x_1, \ldots, x_n$:

$$(16.2.2) \quad L(x_1, \ldots, x_n;\, m_1, \ldots, m_k, \sigma) = \frac{\sqrt{p_1 \cdots p_n}}{\sigma^n (2\pi)^{n/2}} e^{-\frac{1}{2\sigma^2} \sum_1^n p_i (x_i - f_i)^2}.$$

The quantities f_i are here, according to 16.2.1, linear functions of the parameters m_i, and it is required to determine σ and the m_i so as to maximize L. Now it is obvious that for any fixed σ we shall maximize L by minimizing the expression

$$U = \sum_1^n p_i (x_i - f_i)^2.$$

The difference $x_i - f_i$ is the *error* affecting the measurement x_i, and we thus find that according to the maximum likelihood method *the estimates of the unknown true values m_i should be chosen so as to minimize the weighted sum of the squared errors.* — This condition is, in reality, nothing but a straightforward application of the least squares principle (cf. 9.2), and may serve as the starting point for the construction of a theory of errors which is independent of the assumption that errors are normally distributed.

In order to determine the m_i so as to minimize U, we have to substitute for f_i the expressions 16.2.1, and then find values of the m_i such that the partial derivative of U with respect to each m_i reduces to zero. We have

$$\frac{1}{2} \frac{\partial U}{\partial m_r} = m_1 \sum_{i=1}^n p_i a_{ir} a_{i1} + \cdots + m_k \sum_{i=1}^n p_i a_{ir} a_{ik} - \sum_{i=1}^n p_i a_{ir} x_i.$$

Introducing the notations

$$(16.2.3) \qquad b_{rs} = \sum_{i=1}^n p_i a_{ir} a_{is}, \quad c_r = \sum_{i=1}^n p_i a_{ir} x_i,$$

where r and s may take all the values $1, 2, \ldots, k$, we thus obtain the equations

$$b_{11} m_1 + b_{12} m_2 + \cdots + b_{1k} m_k = c_1,$$

$(16.2.4)$ $\qquad \cdot \ \cdot \ \cdot \ \cdot \ \cdot \ \cdot \ \cdot \ \cdot \ \cdot \ \cdot \ \cdot$

$$b_{k1} m_1 + b_{k2} m_2 + \cdots + b_{kk} m_k = c_k,$$

which are known as the *normal equations* connected with the problem.

The symmetric determinant

$$B = \begin{vmatrix} b_{11} & b_{12} & \ldots & b_{1k} \\ . & . & . & . & . \\ b_{k\,1} & b_{k\,2} & \ldots & b_{k\,k} \end{vmatrix}$$

is, according to the theory of determinants, equal to the sum of all the $\binom{n}{k}$ expressions of the form

$$p_{i_1} \cdots p_{i_k} \begin{vmatrix} a_{i_1,1} & \cdots & a_{i_1,k} \\ . & . & . & . \\ a_{i_k 1} & \cdots & a_{i_k k} \end{vmatrix}^2$$

that can be formed by selecting, from among the numbers $1, 2, \ldots, n$, any group of k different numbers $i_1, \ldots, i_k$. It then follows from our hypothesis concerning the functions f_i that we must always have $B > 0$, so that the linear system 16.2.4 has a unique solution. The uniquely determined set of values of the m_i, say $m_1^*, \ldots, m_k^*$, satisfying 16.2.4 will be the required maximum likelihood estimates for $m_1, \ldots, m_k$. From the theory of linear equations we obtain the expressions

(16.2.5)
$$m_s^* = \frac{1}{B} \sum_{r=1}^{k} B_{rs} c_r$$

$$= \frac{1}{B} \sum_{i=1}^{n} p_i x_i \sum_{r=1}^{k} a_{ir} B_{rs},$$

for $s = 1, 2, \ldots, k$, where B_{rs} is the cofactor corresponding to the element b_{rs} in the determinant B. In these expressions, the weights p_i, the coefficients a_{ir}, and the determinants B and B_{rs} are all known quantities, so that the estimate m_s^* will be a linear function of the measurements x_i, with known constant coefficients. By assumption the x_i are independent and normally distributed random variables, and it follows that the m_s^* are also normally distributed. From $E(x_i) = f_i$ we further obtain, using 16.2.1 and 16.2.3,

$$E(m_s^*) = \frac{1}{B} \sum_{i=1}^{n} p_i f_i \sum_{r=1}^{k} a_{ir} B_{rs}$$

$$= \frac{1}{B} \sum_{i=1}^{n} p_i \sum_{t=1}^{k} a_{it} m_t \sum_{r=1}^{k} a_{ir} B_{rs}$$

$$= \frac{1}{B} \sum_{t=1}^{k} m_t \sum_{r=1}^{k} b_{rt} B_{rs}.$$

However, according to the theory of determinants we have

$$\sum_{r=1}^{k} b_{rt} B_{rs} = \begin{cases} B & \text{for} \quad t=s, \\ 0 & \text{for} \quad t \neq s, \end{cases}$$

so that we obtain

(16.2.6) $$E(m_s^*) = m_s.$$

It thus follows that m_s^* is an unbiased estimate (cf. 14.2) of the unknown true value m_s. We further have

$$m_s^* - m_s = \frac{1}{B} \sum_{i=1}^{n} p_i (x_i - f_i) \sum_{r=1}^{k} a_{ir} B_{rs}.$$

Since the x_i are independent variables, and $D^2(x_i) = \sigma^2/p_i$, this gives us after some reductions

(16.2.7) $$D^2(m_s^*) = E((m_s^* - m_s)^2) = \frac{\sigma^2}{B^2} \sum_{r=1}^{k} B_{rs} \sum_{q=1}^{k} b_{qr} B_{qs} = \frac{B_{ss}}{B} \sigma^2.$$

We have thus seen that the maximum likelihood estimate m_s^* is a normally distributed variable, with mean and variance given by 16.2.6 and 16.2.7.

In order to deduce a maximum likelihood estimate also for the unknown parameter σ^2, we obtain from 16.2.2 the equation

$$\frac{\partial \log L}{\partial \sigma} = \frac{1}{\sigma^3} \sum_{i=1}^{n} p_i (x_i - f_i)^2 - \frac{n}{\sigma} = 0,$$

the solution of which is

(16.2.8) $$\sigma^{*2} = \frac{1}{n} \sum_{i=1}^{n} p_i (x_i - f_i^*)^2,$$

where f_i^* denotes the expression

(16.2.9) $$f_i^* = a_{i1} m_1^* + \cdots + a_{ik} m_k^*$$

obtained by replacing, in 16.2.1, each m_i by the corresponding estimate m_i^*.

In the same way as in the last problem of the preceding section, we can now use the argument of 13.4 to show that m_s^* and σ^* are independent, and that $n\sigma^{*2}/\sigma^2$ has a χ^2 distribution with $n-k$ degrees of freedom. In the same way as before it then follows that

$$t = \sqrt{\frac{n-k}{n} \cdot \frac{B}{B_{ss}}} \cdot \frac{m_s^* - m_s}{\sigma^*}$$

has a STUDENT distribution with $n-k$ degrees of freedom. For the true value m_s, we obtain the confidence limits

(16.2.10) $$m_s^* \pm t_p \, \sigma^* \sqrt{\frac{n}{n-k} \cdot \frac{B_{ss}}{B}},$$

where t_p has $n-k$ degrees of freedom. The result of the measurements, as far as m_s is concerned, should be given in the same conventional form as before:

$$m_s^* \pm \sigma^* \sqrt{\frac{n}{n-k} \cdot \frac{B_{ss}}{B}}.$$

The random variables $m_1^*, \ldots, m_k^*$ are linear functions of the independent and normally distributed variables $x_1, \ldots, x_n$. According to 10.4 it follows that the variables m_s^* have a joint normal distribution. The means and variances of this distribution are given by 16.2.6 and 16.2.7, and by an appropriate modification of the calculations leading to 16.2.7 we find the more general formula

(16.2.11) $$E\left((m_r^* - m_r)(m_s^* - m_s)\right) = \frac{B_{rs}}{B} \, \sigma^2.$$

According to 10.4, the joint normal distribution of the variables m_s^* is completely determined by the means 16.2.6 and the second order central moments 16.2.11. In particular it follows that the two variables m_r^* and m_s^* will be independent if, and only if, we have $B_{rs} = 0$.

In many cases it will be required to estimate some given linear function of the true values $m_1, \ldots, m_k$, say the function

$$m = A_1 m_1 + \cdots + A_k m_k,$$

where $A_1, \ldots, A_k$ are given constants. In the same way as above, we obtain for the unknown true value m the confidence limits

(16.2.12) $$A_1 m_1^* + \cdots + A_k m_k^* \pm t_p \, \sigma^* \sqrt{\frac{n}{n-k} \cdot \frac{1}{B} \sum_{r,s=1}^{k} A_r A_s B_{rs}},$$

where t_p still has $n-k$ degrees of freedom.

For the numerical computation of estimates by means of a set of observed values x_i of the functions f_i given by 16.2.1, we first have to find the coefficients b_{rs} and c_r from 16.2.3. and compute the determinants

239

B and B_{rs} formed by the b_{rs}. From 16.2.5 we then obtain the estimates $m_1^*, \ldots, m_k^*$. After having determined σ^* from 16.2.8 and 16.2.9, we can finally compute confidence limits from 16.2.10 and 16.2.12. The quantity of numerical work involved in these operations will be rapidly increasing with the number of the unknown true values $m_1, \cdots, m_k$.[1]

Ex. 1. — From the centre 0 of a circle one sees, turning in the positive direction, the points A, B, C, D on the perimeter, in the order indicated. It is required to determine the angles AOB, BOC, and COD, the unknown true values of which will be denoted by m_1, m_2, and m_3. In order to estimate the m_s, each of the three angles has been measured directly, and measurements have also been performed on the angles AOC and BOD, the true values of which are $m_1 + m_2$ and $m_2 + m_3$ respectively. (The reader should draw a diagram, and work through the numerical computations indicated below.)

For each of the five observed angles a first estimate has been computed as the arithmetical mean of a certain number of measurements, all of which are assumed to have the same degree of precision. The number of measurements for each angle was 8, except in the case of the angle BOC, where 16 measurements were made. According to the preceding section, the weight of each arithmetic mean will be the sum of the weights of the measurements on which the mean is based. Since in our case all measurements have the same weight, and the weights are only determined up to a factor of proportionality, the observed mean for the angle BOC can be given the weight 2, while the corresponding mean for each of the other angles receives the weight 1.

We thus have $k = 3$ unknown true values m_s, and $n = 5$ observed linear functions, as follows:

$$f_1 = m_1,$$
$$f_2 = m_2,$$
$$f_3 = m_3,$$
$$f_4 = m_1 + m_2,$$
$$f_5 = m_2 + m_3.$$

If we regard the arithmetic means mentioned above as the observed values of the f_i, there is just one observed value x_i for each f_i, and the weight of each x_i will be 1, except in the case of x_2, which has the weight 2. The following values were observed:

$$x_1 = 38°\ 59'\ 38''.97,$$
$$x_2 = 59°\ 53'\ 44''.03,$$
$$x_3 = 211°\ 23'\ 46''.38,$$
$$x_4 = 98°\ 53'\ 23''.41,$$
$$x_5 = 271°\ 17'\ 29''.38.$$

---

[1] The theoretical developments of this section could have been considerably simplified by the use of matrix algebra. Since the reader is not assumed to be familiar with this mathematical tool it has, however, not been used here.

From 16.2.3, the coefficients b_{rs} of the system 16.2.4 of normal equations will be found to form the matrix

$$
\begin{matrix}
2 & 1 & 0 \\
1 & 4 & 1 \\
0 & 1 & 2
\end{matrix}
$$

the determinant of which is $B = 12$. The cofactors B_{rs} form the matrix

$$
\begin{matrix}
7 & -2 & 1 \\
-2 & 4 & -2 \\
1 & -2 & 7
\end{matrix}
$$

From 16.2.5 we then obtain the estimates

$$m_1^* = \frac{1}{12}(7x_1 - 4x_2 + x_3 + 5x_4 - x_5) = 38° \ 59' \ 39''.23,$$

$$m_2^* = \frac{1}{12}(-2x_1 + 8x_2 - 2x_3 + 2x_4 + 2x_5) = 59° \ 53' \ 43''.93,$$

$$m_3^* = \frac{1}{12}(x_1 - 4x_2 + 7x_3 - x_4 + 5x_5) = 211° \ 23' \ 45''.92.$$

From 16.2.8 and 16.2.9 we further obtain $\sigma^* = 0''.341$, and hence

$$\sigma^* \sqrt{\frac{n}{n-k} \cdot \frac{{}^1 B_{11}}{B}} = \sigma^* \sqrt{\frac{n}{n-k} \cdot \frac{B_{33}}{B}} = 0''.41,$$

$$\sigma^* \sqrt{\frac{n}{n-k} \cdot \frac{B_{22}}{B}} = 0''.31.$$

The final estimates of the angles m_1, m_2, and m_3 are then, in the conventional form used above,

$$38° \ 59' \ 39''.23 \pm 0''.41,$$
$$59° \ 53' \ 43''.93 \pm 0''.31,$$
$$211° \ 23' \ 45''.92 \pm 0''.41.$$

If it is desired to estimate also, say, the angle AOD, the true value of which is $m_1 + m_2 + m_3$, we obtain from 16.2.12 the result $310° \ 17' \ 9''.08 \pm 0''.54$.

Ex. 2. — When one or more of the observed values are *non-linear* functions of the unknown parameters, it will as a rule be necessary to approximate these by linear functions, using the linear terms of their developments in Taylor series. However, in some cases it may be possible to find exact solutions of the equations of the likelihood method, as the following example shows.

Let m_1, m_2, and m_3 be the unknown lengths of the sides of a rectangular triangle, where $m_3 = \sqrt{m_1^2 + m_2^2}$. By measurements of the same precision we have obtained the observed values x_1, x_2, and x_3 for the three sides. Regarding m_1 and m_2 as our two unknown parameters, we have thus observed two linear and one non-linear function of these. In order to estimate m_1 and m_2 by the maximum likelihood method, we have to find the values of m_1 and m_2 that minimize

$$U = (x_1 - m_1)^2 + (x_2 - m_2)^2 + (x_3 - \sqrt{m_1^2 + m_2^2})^2.$$

The solution of this problem is found to be

$$m_1^* = x_1 \frac{x_3 + \sqrt{x_1^2 + x_2^2}}{2\sqrt{x_1^2 + x_2^2}},$$

$$m_2^* = x_2 \frac{x_3 + \sqrt{x_1^2 + x_2^2}}{2\sqrt{x_1^2 + x_2^2}},$$

$$m_3^* = \frac{x_3 + \sqrt{x_1^2 + x_2^2}}{2}.$$

16.3. A regression problem.

— As an example of an important class of problems, which appear in a number of different applications we now take up the following question.

Consider k *independent variables* $x_1, \ldots, x_k$, and a *dependent variable* y. The independent variables x_i are not random variables, and we make no fixed assumptions as to how their values are determined in a given observation. On the other hand, the dependent variable y, for every given set of values for $x_1, \ldots, x_k$, is a normally distributed random variable with the mean value

(16.3.1) $$\beta_0 + \beta_1 x_1 + \ldots + \beta_k x_k,$$

and the s.d. σ. The numbers $\beta_0, \beta_1, \ldots, \beta_k$, and σ, are unknown parameters.

Suppose now that n independent observations of all of the variables under consideration have been carried out, so that one has at his disposal the n sets of values

$$y_i, x_{1i}, x_{2i}, \ldots, x_{ki} \qquad (i = 1, 2, \ldots, n).$$

It is required to make an estimate of the unknown coefficients $\beta_0, \beta_1, \ldots, \beta_k$ in 16.3.1 by means of these data.

In the case when $x_1, \ldots, x_k$ are also random variables, the expression 16.3.1 is the right-hand side of the equation of the plane of regression of y on x_1, $\ldots, x_k$ (cf. 10.1), and the coefficients β_r are the regression coefficients belonging to this plane. We use a similar terminology in the present case as well, in that we call 16.3.1 a linear regression function for y with respect to $x_1, \ldots, x_k$, and call the β-coefficients the corresponding regression coefficients.

As a typical example of the applications in which regression problems of this type occur, let us consider first the case in which y denotes the amount of coal consumed in a blast furnace per ton of pig iron produced, and in which the variables $x_1, \ldots, x_k$ are quantities which describe the chemical composition of the ore and possibly other quantities which influence the process, such as temperature in the furnace, and so on. If one assumes that y can be considered, with sufficient accuracy, as a random variable having linear regression with respect to $x_1, \ldots, x_k$, then one can perform a series of experiments in order to determine the regression coefficients, where the values of the x-variables can be chosen arbitrarily to a certain extent, having consideration always to the desirability of having the values spread appropriately over the region of variation which is under study. In this case, the x-variables cannot be considered as random variables; and therefore it is natural to state the problem as we have done above. Similar situations occur often in technological and scientific applications, where the circumstances of an experiment often permit one to choose the independent variables' values more or less arbitrarily.

We obtain an example of another group of applications if we let y denote the quantity of a certain commodity, say milk, which is consumed during a certain period of time within some community. The variables $x_1, \ldots, x_k$ are taken here to be the prices for the commodity in question and for other related commodities (butter, bread, and so on) which might be considered to have an influence on y. (As a matter of fact, one ordinarily considers the *logarithms* of both quantities and prices in dealing with problems of this kind.) In contrast to scientific experiments, we can certainly not choose the x-variables at will, but must take the values which are presented by observation. Nevertheless, these values can ordinarily not be considered in any natural way as being the result of random experiments, and therefore it is most reasonable to consider the x-variables as non-random.

Finally, we mention the important example in which the dependent variable y is a function of the time t, and in which y is taken to be the sum of some simple "trend function" and a random disturbance or noise. Such cases appear often in economic and many other applications. For example, with a linear or parabolic trend, one has

$$y = \beta_0 + \beta_1 t + \ldots + \beta_k t^k + z(t),$$

and with a simple periodic trend, one has

$$y = \beta_0 + \beta \, \cos \, (\lambda t + \mu) + z(t).$$

In both cases, $z(t)$ is assumed to be a normally distributed variable for every given t, having mean value 0, and such that the variables $z(t)$ as observed at different time-points are mutually independent and all have the same standard deviation. In the first case cited, one can set $x_r = t^r$ for $r = 1, 2, \ldots, k$, and in the second one can set $k = 1$, $x_1 = \cos$ $(\lambda t + \mu)$. It is clear that in both cases, we have to do with special cases of the general problem described above. The formulas for estimating the coefficients in a linear, parabolic, or trigonometric trend curve from a series of observed functional values $y_1, \ldots, y_n$ corresponding to time points $t_1, \ldots, t_n$ thus appear as special cases of the general formulas given below (see the example at the end of the present section).

It is easy to see that the problem set forth above is actually identical with the problem in error theory treated in the previous section, aside from differences in notation.

Indeed, let us replace the parameters $\beta_0, \beta_1, \ldots, \beta_k$ in the present problem by the "true values" $m_1, \ldots, m_{k+1}$ for $k+1$ unknown quantities, and the known coefficients $1, x_{1i}, \ldots, x_{ki}$ for the parameters in the i-th observation by the numbers $a_{i1}, \ldots, a_{i, k+1}$. Finally, let us replace the observed values $y_1, \ldots, y_n$ by $x_1, \ldots, x_n$. Under this change of notation, it is plain that the problem of estimating the parameters $\beta_0, \beta_1, \ldots, \beta_k$ goes over into the problem of estimating the $k+1$ unknown true values $m_1, \ldots, m_{k+1}$, under the same hypotheses as in the previous section, with the simplification that all of the weights p_i are equal to unity and that all coefficients for the first unknown m_1 in 16.2.1 are equal to unity.

The present problem is thus solved by the simple procedure of replacing k by $k+1$ in the formulas of the previous section and then carrying out the indicated substitutions. We give here only the result, omitting the details of the calculation. These will be useful for the reader to carry out for himself.

We first introduce the following notation (r and s can take on all of the values $1, 2, \ldots, k$):

$$\bar{x}_r = \frac{1}{n} \sum_{i=1}^{n} x_{ri}, \qquad \bar{y} = \frac{1}{n} \sum_{i=1}^{n} y_i,$$

$$l_{rs} = \frac{1}{n} \sum_{i=1}^{n} (x_{ri} - \bar{x}_r)(x_{si} - \bar{x}_s), \qquad l_{os} = \frac{1}{n} \sum_{i=1}^{n} (y_i - \bar{y})(x_{si} - \bar{x}_s),$$

$$L = \begin{vmatrix} l_{11} & l_{12} & \cdots & l_{1k} \\ \cdot & \cdot & \cdot & \cdot & \cdot \\ l_{k1} & l_{k2} & & l_{kk} \end{vmatrix}.$$

Also, let L_{rs} be the cofactor corresponding to the element l_{rs} in the determinant L.

The estimation formulas become somewhat simpler than they would otherwise be if one makes a slight change in the definitions of the parameters which are to be found. Namely, in 16.3.1, we replace β_0 by $\alpha - \beta_1 \bar{x}_1 - \ldots - \beta_k \bar{x}_k$. For a given system of values $x_1, \ldots, x_k$, then, the mean value for y will be given by the expression

$$a + \beta_1 (x_1 - \bar{x}_1) + \ldots + \beta_k (x_k - \bar{x}_k)$$

instead of by 16.3.1. Our task is now to estimate the parameters α and $\beta_1, \ldots, \beta_k$. Let α^* and $\beta_1^*, \ldots, \beta_k^*$ be the values of these parameters estimated by the maximum likelihood method. Presenting these values with their estimated standard errors in the usual way, we have

$$\alpha^* = \bar{y} \pm \frac{\sigma^*}{\sqrt{n-k-1}},$$

$$\beta_r^* = \frac{1}{L} \sum_{s=1}^{k} l_{os} L_{rs} \pm \frac{\sigma^*}{\sqrt{n-k-1}} \cdot \sqrt{\frac{L_{rr}}{L}},$$

where

$$\sigma^{*2} = \frac{1}{n} \sum_{i=1}^{n} (y_i - a^* - \beta_1^* (x_{1i} - \bar{x}_1) - \cdots - \beta_k^* (x_{ki} - \bar{x}_k))^2.$$

It is often required (as in 16.2.12) also to estimate the mean value of y for a given system $X_1, \ldots, X_k$ of values of the independent variables. We obtain the estimate

$$a^* + \beta_1^* (X_1 - \bar{x}_1) + \cdots + \beta_k^* (X_k - \bar{x}_k) \pm$$

$$\pm \frac{\sigma^*}{\sqrt{n-k-1}} \sqrt{1 + \frac{1}{L} \sum_{r,s=1}^{k} L_{rs} (X_r - \bar{x}_r)(X_s - \bar{x}_s)}.$$

Confidence limits for the estimated values are obtained as before by multiplying the standard errors in the above formulas with the factor t_p, where t_p now has $n - k - 1$ degrees of freedom.

Let α and β be coefficients in a linear trend function (see the above discussion in small print), which we write in the form $\alpha + \beta(t - \bar{t})$. Suppose that the dependent variable y is observed at the time points $t_1, \ldots, t_n$. We set

$$\bar{t} = \frac{1}{n} \sum_{i=1}^{n} t_i, \quad w^2 = \frac{1}{n} \sum_{i=1}^{n} (t_i - \bar{t})^2.$$

For α and β, we obtain the estimates

$$a^* = \bar{y} \pm \frac{\sigma^*}{\sqrt{n-2}},$$

$$\beta^* = \frac{1}{w^2 n} \sum_{i=1}^{n} (t_i - \bar{t})(y_i - \bar{y}) \pm \frac{\sigma^*}{w\sqrt{n-2}},$$

where

$$\sigma^{*2} = \frac{1}{n} \sum_{i=1}^{n} (y_i - a^* - \beta^* (t_i - \bar{t}))^2.$$

The ordinate of the regression line $\alpha + \beta(T - \bar{t})$, computed at a given time point T, is estimated by the following formula:

$$a^* + \beta^* (T - \bar{t}) \pm \frac{\sigma^*}{\sqrt{n-2}} \sqrt{1 + \left(\frac{T-\bar{t}}{w}\right)^2}.$$

16.4. Analysis of variance. — The technique of *analysis of variance*, introduced by R. A. FISHER, has proved to be widely applicable in experimental investigations having to do with comparisons of different methods of treatment, different varieties of plants or races of animals, different qualities of raw material, and so on. This technique has found its widest application up to the present time within the field of biology, and especially in agricultural experimentation. Nevertheless, it can perform important services in other fields as well, for example in industrial studies. Here we can only point out briefly a few of the simplest cases in which analysis of variance can be applied. For completion of the mathematical proofs, the reader is referred to *Mathematical methods*, Chapter 36; for a more exhaustive presentation of the method from the biological standpoint, the reader is referred to the works of FISHER, COCHRAN-COX, and KEMPTHORNE mentioned in the bibliography.

We take up first the following problem. One observes r independent normally distributed variables, all of which are assumed to have one and the same unknown standard deviation σ. The unknown mean values $m_1, \ldots, m_r$ may possibly be different. A series of observations has been carried out for each of the variables, wherein the i-th series contains the n_i observed values $x_{i1}, \ldots, x_{in_i}$. *With this information, it is required to test the hypothesis that all of the mean values are equal:* $m_1 = m_2 = \ldots = m_r$.

For example, the observed variables can be yields per acre of r different strains of wheat, outputs of a certain industrial commodity under r different methods of manufacture, and so on. If the proposed hypothesis is to be accepted, then the data must not show any statistically significant differences among the different strains or methods of manufacture. On the other hand, if such differences are present, the hypothesis must be rejected.

The case $r = 2$ has already been treated in 14.4, example 3. The problem now under consideration is more general, since one wishes to make a simultaneous comparison of an arbitrary number of mean values.

To treat this problem, we set

$$\bar{x}_i = \frac{1}{n_i} \sum_{j=1}^{n_i} x_{ij}, \quad n = \sum_{i=1}^{r} n_i,$$

$$\bar{x} = \frac{1}{n} \sum_{i=1}^{r} \sum_{j=1}^{n_i} x_{ij} = \frac{1}{n} \sum_{i=1}^{r} n_i \bar{x}_i,$$

so that $\bar{x}_i$ and $\bar{x}$, respectively, denote the observed mean values for the i-th series and for all observations. It is then simple to verify the fundamental identity

$$\sum_{i=1}^{r} \sum_{j=1}^{n_i} (x_{ij} - \bar{x})^2 = \sum_{i=1}^{r} n_i (\bar{x}_i - \bar{x})^2 + \sum_{i=1}^{r} \sum_{j=1}^{n_i} (x_{ij} - \bar{x}_i)^2,$$

which we write in abbreviated form as

(16.4.1) $$Q = Q_1 + Q_2.$$

Thus Q_1 denotes the sum of the squares of the differences between the mean values $\bar{x}_i$ and the general mean value $\bar{x}$, weighted with the numbers n_i. Furthermore, Q_2 denotes the sum of the squares of the differences between each individual observation x_{ij} and the corresponding series mean value $\bar{x}_i$. The number Q, finally, is the sum of the squares of the differences between the individual observations and the general mean value. The number Q_1 is called the sum of squares *between series*, Q_2 the sum of squares *within series*, and Q the *total* sum of squares.

Suppose now for a moment that the proposed hypothesis is correct. Then all of the r variables have one and the same normal distribution, and Q/n is simply the observed variance for the material of all n observations formed by all series taken together. According to 13.4, Q/σ^2 then has a χ^2-distribution with $n - 1$ degrees of freedom. Just as in 13.4, it can be shown that Q_1/σ^2 and Q_2/σ^2 have χ^2-distributions with $r - 1$ and $n - r$ degrees of freedom, respectively. Furthermore, Q_1 and Q_2 are independent. According to 8.1, we then have

$$E\left\{\frac{1}{n-1} Q\right\} = E\left\{\frac{1}{r-1} Q_1\right\} = E\left\{\frac{1}{n-r} Q_2\right\} = \sigma^2.$$

Next, 8.3 shows that the quotient of the mean squares

$$F = \frac{\dfrac{1}{r-1} Q_1}{\dfrac{1}{n-r} Q_2}$$

has an F-distribution with $r - 1$ degrees of freedom in the numerator and $n - r$ degrees of freedom in the denominator. If p is a given percentage, and if F_p denotes the $p\%$-value for the F-distribution (see 8.3), then we have a probability of $p\%$ that F assumes a value $> F_p$.

Suppose on the other hand that the hypothesis is false, that is, that not all of the mean values m_i are equal. The sum of squares Q_2 obviously remains unaltered if we replace each x_{ij} by $x_{ij} - m_i$, so that the distribution and mean value of Q_2 are the same as in the former case. On the other hand, it is easy to see that every non-equality among the quantities m_i tends to increase the value of the sum Q_1 and therefore also of the quotient F.

We now introduce the rule that the hypothesis $m_1 = \ldots = m_r$ shall be accepted if $F \leq F_p$ and rejected if $F > F_p$.

This clearly constitutes a test on the $p\,\%$ level (see 14.4—14.6), since one has an *a priori* probability of $p\,\%$ that the hypothesis will be rejected when it is in fact correct. Since any failure of the hypothesis favors a large value of F, we have also taken into consideration the second requirement for a criterion for testing a statistical hypothesis, namely, that it should make the probability of rejecting a false hypothesis as large as possible (see 14.6).

Ex. 1. — Suppose that the lifetime of an electric bulb is a normally distributed variable, and that moderate changes in methods of manufacture or raw material can affect the mean value of the distribution but not its standard deviation (see the examples in 14.4). A number of bulbs were taken at random from four different lots, and the following lifetimes in hours were obtained.

Lot 1:	1600,	1610,	1650,	1680,	1700,	1720,	1800,	
Lot 2:	1580,	1640,	1640,	1700,	1750,			
Lot 3:	1460,	1550,	1600,	1620,	1640,	1660,	1740,	1820,
Lot 4:	1510,	1520,	1530,	1570,	1600,	1680.		

The mean values for the four series, each rounded off to an integer number of hours, are 1680, 1662, 1636, and 1568, in that order. In order to test how far the non-equalities between these values are to be considered as significant, one must compute the sums Q_1 and Q_2 and the quotient F. The numbers obtained in the computation are usually arranged in the following tabular form.

Type of variation	Degrees of freedom	Sum of squares	Mean square
Between lots	$r - 1 = 3$	$Q_1 = 44{,}361$	14,787
Within lots	$n - r = 22$	$Q_2 = 151{,}351$	6,880
Total	$n - 1 = 25$	$Q = 195{,}712$	

We find that $F = 14{,}787/6{,}880 = 2.15$. Since this value is under the 5%-value according to Table V, which is 3.05, the differences between the mean values cannot be considered as significant, but the data must be judged to be consistent with the hypothesis that the four lots have the same mean lifetime.

The observations x_{ij} in the foregoing problem admitted only one cause of variation, namely, their belonging to one or another of the r different series. Within a single series, it was assumed that all observations have the same distribution, so that the non-equalities among them all have random causes.

The greatest importance of the analysis of variance, however, lies in the fact that with it one can simultaneously investigate several causes of variation. In an agricultural experiment, for example, it may be important to take account of variations in yield between various experimental plots which depend upon different plant strains, different fertilizers, and differences in fertility of the soil. In an industrial investigation, it may be a question of the effect upon the product of different machines, different raw materials, and the like.

Treatment of different problems by analysis of variance rests upon an extension of the identity 16.4.1. The total sum of squares Q for all observations is divided up into a number of components, and one tries to arrange this division in such a fashion that every actual or suspected cause of variation is expressed in a fixed component. In this way, the effect of the various causes are isolated from each other and can be studied separately.

Suppose that one wishes to compare the yields of r types of a certain kind of seed. One lays out a certain number of plots of equal size in a field, which are sown with the different types of seed. One then determines the yield for each plot. If the fertility of the soil could be taken as constant over the entire field, then the mean values for the yields of the different types of seed could be compared by the method set forth above. Ordinarily, however, such an hypothesis is not even approximately correct, for the fertility of the soil varies noticeably in different part of the fields. The size of the yield for different plots therefore is subject to the two essential causes of variation, "type" and "fertility". One wishes to isolate these from each other in order to be able to study possible variations between the types without being confused by variations in fertility.[1]

One can proceed in the following way. In different parts of the field, one lays out an appropriate number s of connected "blocks", each of which should be as homogeneous as possible with respect to fertility. Each block is divided into r plots of equal size, and each plot is then sown with one of the r types of seed, in an order which should be determined by chance. Let x_{ij} denote the size of the yield in the plot with the i-th type in the j-th block. The value assumed by the variable x_{ij} in one experiment is influenced partly by the two essential causes of variation, type and fertility, and partly by a large number of causes which we assume to be random. The variation in fertility within each block can be considered to be one of these random causes,

---

[1] Although the interest in the present example is primarily in one of the causes of variation, the same scheme can be used in other cases to study simultaneously both causes of variation.

since the order of sowing within each block has been "randomized". It is now natural to assume that the two essential causes of variation have their expression in a "type effect" and a "block effect", which are combined additively. To the sum of these effects, there is added a normally distributed error which is the effect of the random causes. Each x_{ij} should then be capable of being written in the form

$$(16.4.2) \qquad x_{ij} = \alpha + \beta_i + \gamma_j + \xi_{ij},$$

where α, β_i, and γ_j are constants (β_i represents the type effect and γ_j the block effect), while ξ_{ij} is a normally distributed variable with mean value 0. There is cearly no loss of generality in assuming that

$$\sum_{i=1}^{r} \beta_i = \sum_{j=1}^{s} \gamma_j = 0,$$

since this condition can always be satisfied by an appropriate change of the constant α. We make the final assumption that the variables ξ_{ij} are independent and have the same standard deviation σ.

We now wish to test the hypothesis that all of the types have the same value, that is, that $\beta_1 = \ldots = \beta_r = 0$. We set

$$\bar{x}_i. = \frac{1}{s} \sum_{j=1}^{s} x_{ij}, \quad \bar{x}._j = \frac{1}{r} \sum_{i=1}^{r} x_{ij}, \quad \bar{x} = \frac{1}{rs} \sum_{i=1}^{r} \sum_{j=1}^{s} x_{ij},$$

so that $\bar{x}_i.$ denotes the mean yield for the i-th type, and similarly $\bar{x}._j$ for the j-th block, while $\bar{x}$ is the mean yield for all of the rs plots. In analogy with 16.4.1, we can now divide up to the total sum of squares into three components:

$$\sum_{i=1}^{r} \sum_{j=1}^{s} (x_{ij} - \bar{x})^2 = s \sum_{i=1}^{r} (\bar{x}_i. - \bar{x})^2 + r \sum_{j=1}^{s} (\bar{x}._j - \bar{x})^2 + \sum_{i=1}^{r} \sum_{j=1}^{s} (x_{ij} - \bar{x}_i. - \bar{x}._j + \bar{x})^2$$

or

$$(16.4.3) \qquad Q = Q_1 + Q_2 + Q_3,$$

where Q_1 and Q_2 are the square sums *between types* and *between blocks*, respectively. For reasons which will be made apparent below, Q_3 is called the sum of squares for the *error*.

If the hypothesis $\beta_1 = \ldots = \beta_r = 0$ is correct, it can be shown that the square sums Q_1 and Q_3 are independent, no matter what the block effects γ_j may be. After division by σ^2, the sums Q_1 and Q_3 have a χ^2-distribution with

$r-1$ and $(r-1)(s-1)$ degrees of freedom, respectively. We therefore obtain

$$E\left\{\frac{1}{r-1}Q_1\right\} = E\left\{\frac{1}{(r-1)(s-1)}Q_3\right\} = \sigma^2.$$

Further, the quotient of the mean squares,

$$F = \frac{\dfrac{1}{r-1}Q_1}{\dfrac{1}{(r-1)(s-1)}Q_3},$$

has an F-distribution with $r-1$ degrees of freedom in the numerator and $(r-1)(s-1)$ in the denominator.

On the other hand, if the hypothesis is false, it is shown just as in the previous problem that Q_3 still has the same distribution and the same mean value as before. Whether the hypothesis is correct or incorrect, the number Q_3 divided by the number of degrees of freedom thus gives an estimate with the correct mean value for σ^2. This motivates the name for Q_3 introduced above. On the other hand, any incorrectness in the hypothesis tends to increase Q_1 and hence the quotient F as well.

Just as before, we now obtain a test on the p % level: accept the hypothesis if $F \leq F_p$ and reject it if $F > F_p$.

It is often of interest as well to give a direct estimate of the unknown difference $\beta_i - \beta_j$ between effects of two different types. For this difference, we have the confidence limits

$$\bar{x}_i. - \bar{x}_j. \pm t_p \sqrt{\frac{2}{s} \cdot \frac{Q_3}{(r-1)(s-1)}},$$

where t_p is taken from Table IV with $(r-1)(s-1)$ degrees of freedom. For the formula to hold, however, it is necessary that the two types compared be chosen in advance and not selected on the basis of the result of the experiment. For example, one may not select for comparison the two types which gave in the experiment the largest and the smallest mean yield.

It follows from the expression 16.4.2 that the probability scheme used here is a special case of the general regression scheme described in 16.3. The regression coefficients are represented here by the constant α, β_i, and γ_j, while the variables x in the present context can assume only the values 0 and 1.

Ex. 2. — In an agricultural experiment with $r = 6$ types and $s = 5$ blocks, the following figures were obtained for the yield of grain in kilograms per plot of 40 square meters.

Block number	Type number						Total
	1	2	3	4	5	6	
1	12.0	11.5	11.5	11.0	9.5	9.3	64.8
2	10.8	11.4	12.0	11.1	9.6	9.7	64.6
3	13.2	13.1	12.5	11.4	12.4	10.4	73.0
4	14.0	14.0	14.0	12.3	11.5	9.5	75.3
5	14.6	13.2	14.2	14.3	13.7	12.0	82.0
Total	64.6	63.2	64.2	60.1	56.7	50.9	359.7

Type no. 1 is taken here to be a "gauge type", whose properties are known in advance. The other types are to be tested. It is first required to decide whether or not there exist significant differences among the types. Furthermore, it is desired to compare type 1 with types 2 and 5. Analysis of variance gives the following figures, in the computation of which it is useful to apply the formulas

$$Q = \sum_{i=1}^{r} \sum_{j=1}^{s} x_{ij}^2 - \frac{1}{rs} \left(\sum_{i=1}^{r} \sum_{j=1}^{s} x_{ij} \right)^2 ,$$

$$Q_1 = \frac{1}{s} \sum_{i=1}^{r} \left(\sum_{j=1}^{s} x_{ij} \right)^2 - \frac{1}{rs} \left(\sum_{i=1}^{r} \sum_{j=1}^{s} x_{ij} \right)^2 ,$$

$$Q_2 = \frac{1}{r} \sum_{j=1}^{s} \left(\sum_{i=1}^{r} x_{ij} \right)^2 - \frac{1}{rs} \left(\sum_{i=1}^{r} \sum_{j=1}^{s} x_{ij} \right)^2 ,$$

and then to find Q_3 by subtraction, according to 16.4.3.

Cause of variation	Degrees of freedom	Sum of squares	Mean square
Between types.	$r-1=5$	$Q_1 = 28.55$	5.710
Between blocks	$s-1=4$	$Q_2 = 36.41$	9.102
Error	$(r-1)(s-1)=20$	$Q_3 = 9.53$	0.476
Total	$rs-1=29$	$Q = 74.49$	

From this, we find $F = 5.710/0.476 = 12.00$. According to Table VI, the corresponding 1%-value is only 4.10, so that the differences between types are significant.

For comparison of the types 1 and 2, one has the t-value

$$\frac{\bar{x}_1. - \bar{x}_2.}{\sqrt{\frac{2}{s} \cdot \frac{Q_3}{(r-1)(s-1)}}} = 0.642,$$

which for 20 degrees of freedom lies between the 50 and the 60% values according to Table IV. Hence no significant difference can be detected between these two types.

For types 1 and 5, the value $t = 3.621$ is obtained, however. This is greater than the 1%-value according to Table IV, so that the difference in this case is significant.

16.5. Sampling. — A variable X is supposed to have a determined value for each of the N individuals in a given population. Numbering the individuals in an arbitrary order, we denote these values by $x_1, \ldots, x_N$ and set

$$m = \frac{1}{N} \sum_{\nu=1}^{N} x_\nu, \quad \sigma^2 = \frac{1}{N} \sum_{\nu=1}^{N} (x_\nu - m)^2.$$

The values x, as well as the mean value m and the variance σ^2, are supposed to be unknown. A sample of n individuals is chosen at random from the population, and the values of the variable are observed for the individuals in this sample: denote these values by $x_{\nu_1}, \ldots, x_{\nu_n}$. It is required to estimate the unknown mean value m for the whole population from these sample values.

In most of the practical applications where this problem appears, one is concerned with *sampling without replacement* (see 4.2), and here we shall treat only this case. We suppose furthermore that the choice is made by purely random means, so that every individual in the population has the same probability of being selected in a sample (see also the section in small print on page 256—257).

We use the customary notation for the mean value and the variance of the sample:

$$\bar{x} = \frac{1}{n} \sum_{i=1}^{n} x_{\nu_i}, \quad s^2 = \frac{1}{n} \sum_{i=1}^{n} (x_{\nu_i} - \bar{x})^2.$$

When a sample is taken, it is obvious that the values of $\bar{x}$ and s^2 depend upon the individuals which appear in the sample. Therefore $\bar{x}$ and s^2 are random variables, and we shall now compute the mean values of these variables, as well as the variance of $\bar{x}$.

For every $\nu = 1, 2, .., N$, we introduce a random variable ε_ν, which assumes the value 1 if the individual with number ν appears in the sample and assumes the value 0 if the individual with number ν does not appear in the sample. For every fixed ν, we have, according to the fundamental formula 4.2.1,

$$P(\varepsilon_\nu = 1) = \frac{\binom{N-1}{n-1}}{\binom{N}{n}} = \frac{n}{N},$$

and for $\mu \neq \nu$,

$$P\left(\varepsilon_\mu=\varepsilon_\nu=1\right)=\frac{\binom{N-2}{n-2}}{\binom{N}{n}}=\frac{n\,(n-1)}{N\,(N-1)}\,.$$

It follows that

$$E\left(\varepsilon_\nu\right)=E\left(\varepsilon_\nu^2\right)=\frac{n}{N},\quad E\left(\varepsilon_\mu\,\varepsilon_\nu\right)=\frac{n\,(n-1)}{N\,(N-1)}\,.$$

Now, the expressions for $\bar{x}$ and s^2 can be written

$$\bar{x}=\frac{1}{n}\sum_{\nu=1}^{N}\varepsilon_\nu\,x_\nu,\quad s^2=\frac{1}{n}\sum_{\nu=1}^{N}\varepsilon_\nu\,(x_\nu-\bar{x})^2=\frac{1}{n}\sum_{\nu=1}^{N}\varepsilon_\nu\,x_\nu^2-\bar{x}^2,$$

and one finds at once that

$$E\left(\bar{x}\right)=\frac{1}{n}\sum_{\nu=1}^{N}E\left(\varepsilon_\nu\right)x_\nu=\frac{1}{N}\sum_{\nu=1}^{N}x_\nu=m.$$

After a little calculation (the reader should verify the formulas used!), we obtain

$$D^2\left(\bar{x}\right)=E\left(\bar{x}^2\right)-E^2\left(\bar{x}\right)$$

$$=\frac{1}{n^2}\sum_{\mu=1}^{N}\sum_{\nu=1}^{N}E\left(\varepsilon_\mu\,\varepsilon_\nu\right)x_\mu\,x_\nu-m^2$$

(16.5.1)
$$=\frac{N-n}{N-1}\cdot\frac{\sigma^2}{n}\,,$$

$$E\left(s^2\right)=\frac{1}{N}\sum_{\nu=1}^{N}x_\nu^2-E\left(\bar{x}^2\right)=\frac{N}{N-1}\cdot\frac{n-1}{n}\,\sigma^2.$$

The sample mean $\bar{x}$ can thus be used as an estimate for the unknown mean value m of the entire population. The last formula gives us the estimate $\dfrac{N-1}{N}\dfrac{n}{n-1}s^2$, for the variance σ^2. Upon substituting this value for σ in the formula for $D^2\left(\bar{x}\right)$, we obtain the following estimate for m, with separate standard error term:

(16.5.2)
$$\bar{x}\pm s\sqrt{\frac{N-n}{N\,(n-1)}}\,.$$

Ex. — In the special case where a certain number Np of the values x_ν are equal to 1 and the remaining $N(1-p)=Nq$ are equal to 0, we have $m=p$ and $\sigma^2=pq$. Hence

$$E\left(\bar{x}\right)=p, \quad D^2\left(\bar{x}\right)=\frac{N-n}{N-1}\cdot\frac{p\,q}{n}\,.$$

In this case, $\bar{x}$ can be interpreted as the relative frequency of white balls in a sample of n balls which are drawn without replacement from an urn which contains Np white balls and Nq black balls (cf. 5.11, ex. 8).

Suppose now that the given population of N individuals can be divided up into r sub-populations containing $N_1, \ldots, N_r$ individuals, where the numbers N_i are known. We set

$$N_i = N\,p_i,$$

so that

$$\sum_{i=1}^{r} N_i = N, \quad \sum_{i=1}^{r} p_i = 1.$$

The mean value and the variance of the X-values in the i-th sub-population are denoted by m_i and σ_i^2 respectively. These values, like m and σ^2 previously, are supposed to be unknown. It is easy to see that

$$m = \sum_{i=1}^{r} p_i\, m_i$$

and that

(16.5.3) $$\sigma^2 = \sum_{i=1}^{r} p_i\left(\sigma_i^2 + (m_i - m)^2\right).$$

To estimate the mean value m on the basis of a sample of n individuals one can choose a sample directly from the total population, just as in the previous case, without distinguishing among the sub-populations. Just as before, one obtains the estimated value $\bar{x}$ with its mean error, as given in 16.5.2. However, it is natural to see if it is possible to obtain a closer estimate of m by employing a *stratified* sample. Thus one chooses a sample from each of the sub-populations in such a way that the r samples contain n individuals all together, and then estimates m by an appropriate combination of the results obtained so far. We shall consider here only the so-called *representative* method of sampling, in which the number of individuals in the various samples are taken to be proportional to the size of the respective sub-populations or *strata*.

Thus $n_i = n\,p_i$ individuals are taken at random from the i-th stratum. Let $\bar{x}_i$ and s_i^2 denote the mean value and the variance for the observed X-values in this sample. Just as before, one has

$$E\left(\bar{x}_i\right)=m_i, \quad D^2\left(\bar{x}_i\right)=\frac{N_i-n_i}{N_i-1}\cdot\frac{\sigma_i^2}{n_i}, \quad E\left(s_i^2\right)=\frac{N_i}{N_i-1}\cdot\frac{n_i-1}{n_i}\,\sigma_i^2.$$

If we now write

$$\bar{\bar{x}} = \sum_{i=1}^{r} p_i \, \bar{x}_i,$$

then we have

$$E\left(\bar{\bar{x}}\right) = \sum_{i=1}^{r} p_i \, m_i = m,$$

(16.5.4)

$$D^2\left(\bar{\bar{x}}\right) = \sum_{i=1}^{r} p_i^2 \frac{N_i - n_i}{N_i - 1} \cdot \frac{\sigma_i^2}{n_i} = \frac{N - n}{n} \sum_{i=1}^{r} \frac{p_i^2 \, \sigma_i^2}{N \, p_i - 1}.$$

Now replace σ_i^2 by the estimated value $\dfrac{N_i - 1}{N_i} \cdot \dfrac{n_i}{n_i - 1} s_i^2$. One obtains the following estimate for m, with separate standard error term:

$$\bar{\bar{x}} \pm \sqrt{\frac{N - n}{N} \sum_{i=1}^{r} \frac{p_i^2 \, s_i^2}{n \, p_i - 1}}.$$

To compare the accuracy of the two estimates $\bar{x}$ and $\bar{\bar{x}}$, we form the difference between their variances. From 16.5.1, 16.5.3, and 16.5.4 we obtain, after some computation, the result

$$D^2\left(\bar{x}\right) - D^2\left(\bar{\bar{x}}\right) = \frac{N - n}{n\left(N - 1\right)} \left(\sum_{i=1}^{r} p_i \left(m_i - m\right)^2 - \sum_{i=1}^{r} \frac{p_i \left(1 - p_i\right)}{N \, p_i - 1} \sigma_i^2 \right).$$

For given values of p_i and σ_i, the negative terms inside the parentheses go to zero as $N \to \infty$. Under the assumption that not all of the mean values m_i are equal, the positive terms will dominate the negative terms for sufficiently large N, so that the difference will be positive. Under the conditions met with in practice, N is usually so large that this condition is satisfied, and the stratified sample gives an estimate of greater accuracy than the overall sample.

Stratified samples have important applications in public opinion polls, demographic and social investigations, and the like. In such applications, however, one often encounters sampling which is to a greater or lesser extent non-random. A detailed account of the questions which arise in this connection is outside the scope of this work.

In practice, a full realization of randomness in taking a sample is often fraught with serious difficulties. To avoid an unwarranted "favoring" of certain groups or individuals, the process of selection must be made as objective as possible. In certain cases, one can profitably employ tables of so-called *random numbers*. These tables contain columns of digits which can be considered as series of observed values of a variable which assumes each of the values 0, 1, ..., 9 with equal probability. To take a sample of a given population with the aid of such a table, one first numbers the individuals in the population in an arbitrary order. Each number used can always be taken to consist of the same number of digits. For example, if the number of individuals is greater than 100 and less than 1000, one can use the numbers 000–999. Then one reads from the table a series of numbers with the same number of digits, and the individuals with these numbers are included in the sample. One continues in this way, ignoring numbers in the table with no individual corresponding to them, as well as numbers previously used, until the specified number of individuals has been selected. Such tables can be found, for example, in the work by FISHER and YATES mentioned in the bibliography.

16.6. Statistical quality control. — A special application of sampling techniques, which has recently become very important, is *quality control* in industry. Such control is applied both in connection with finished goods and at various stages in the manufacturing process. Here we shall discuss only the first-named case.

Suppose that a commodity is produced in lots of N similar units, which are to satisfy certain specified requirements, such as for example that certain dimensions shall lie between given limits, that strength shall be above certain minimum values, and so on. Suppose further that a unit taken out for testing can always be classified unambiguously as being *satisfactory* or *defective* with respect to the requirements laid down.

Certain units from each lot are taken out for testing. The rules for selection and for deciding, on the basis of the test, whether the lot shall be accepted or handled on a different basis (rejected, submitted to other tests, and so on), comprise the *sampling plan*.

Several different types of such plans are met with in practice. Here we shall consider only one of the simplest types, the so-called process of *single sampling*. This plan has the following provisos. From each lot of N units a sample of n units is taken by random selection. These units are tested. If at most r defective units are found, then the lot is approved. If more than r defective units are found, then all of the units in the lot are tested.[1] In both cases, the manufacturer is obliged to replace all defective units which are found with satisfactory ones.

Suppose that the true (and unknown) number of defective units in the lot is Np. From 4.2.1, it follows that the probability that the lot will be approved without testing all of the units is equal to

$$(16.6.1) \qquad f(p) = \sum_{i=0}^{r} \frac{\binom{Np}{i}\binom{Nq}{n-i}}{\binom{N}{n}},$$

---

[1] In cases where *destructive testing* is used, this proviso must naturally be modified.

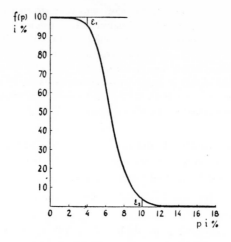

Fig. 29. *OC*-curve for single sampling. $N = 3400$, $n = 225$, $r = 14$. For the marked points, $p_1 = 4\ \%$, $p_2 = 10\ \%$, and, approximately, $\varepsilon_1 = \varepsilon_2 = 4\ \%$. For this scheme, *AOQL* is 4.2 %.

where $q = 1 - p$. The curve which represents the function $y = f(p)$ is usually called the plan's *OC*-curve (operating characteristic curve). The function $f(p)$ is non-increasing as p goes from 0 to 1, and also $f(0) = 1$, $f(1) = 0$. The general appearance of an *OC*-curve is therefore as sketched in Figure 29. (The reader should prove this.) Of course, the function $f(p)$ is defined by 16.6.1 only for values of p such that Np is an integer, but it can be extended by interpolation to all values of p in the interval $0 < p < 1$ if the binomial coefficients are expressed in terms of the Γ-function.

If N is large in comparison with n, then, as shown in 4.2, the function $f(p)$ can be approximated by the corresponding expression for sampling with replacement:

$$\sum_{i=0}^{r} \binom{n}{i} p^i q^{n-i},$$

In practice, this expression often gives sufficient accuracy.

If N is supposed to be given, the plan just given contains two free parameters, n and r. Within certain limits, these parameters can be determined in such a way that the plan has some properties that are considered desirable in a given case. Various properties can come under consideration, and we shall briefly discuss some of them.

Having reference to the requirements imposed upon the product under consideration, it may be possible to determine two numbers p_1 and $p_2 > p_1$, so that p_1 is the largest frequency of defective units which can be considered as consistent with good quality in a lot while p_2 is the largest frequency of defective units which it would be efficient to accept. From the producer's standpoint, it is therefore desirable that the plan should give a high probability for accepting a lot for which $p \leq p_1$, and from the consumer's standpoint it is desirable that there should be a low probability of accepting a lot in which $p \geq p_2$. For example, one may demand that the plan shall satisfy the conditions

$$f(p_1) = 1 - \varepsilon_1,$$

(16.6.2)

$$f(p_2) = \varepsilon_2,$$

where ε_1 and ε_2 are conveniently chosen small positive quantities. ε_1 is called the *producer's risk* and ε_2 is called the *consumer's risk*.

The conditions 16.6.2 imply that the OC-curve must pass through the two points $(p_1, 1 - \varepsilon_1)$ and (p_2, ε_2). In practice, one can usually determine integer values for n and r so that these conditions are at least approximately satisfied. (See Fig. 29.) The required number n of units in the sample will obviously increase in the same degree as p_1 and p_2 are chosen near to each other or the risks ε_1 and ε_2 are made small.

The following reasoning leads to another condition which may be found desirable to impose on a sampling plan. If the true number of defective units in the lot is Np, and if the number of defective units found in the sample is $i \leq r$, then the lot is accepted and contains $Np - i$ defective units, since the defective units detected in the sample have been replaced. On the other hand, if more than r defective units are found in the sample, then the entire lot is tested, and contains after its final acceptance only satisfactory units. The mean value of the relative frequency of defective units in a lot which is accepted after testing and replacement of defective units is thus

$$g(p) = \sum_{i=0}^{r} \left(p - \frac{i}{N} \right) \frac{\binom{Np}{i} \binom{Nq}{n-i}}{\binom{N}{n}}.$$

One sees without difficulty that $g(p) < p$ for all p, $0 < p < 1$. The test therefore has the effect of improving the quality. It is also clear that $g(0) = g(1) = 0$. Hence the function $g(p)$ has a maximum value for a certain value of p between 0 and 1. The value of this maximum depends upon n and r. This maximum value for $g(p)$ is called the plan's $AOQL$ (*average outgoing quality limit*). Regardless of the quality of the incoming lots, the average frequency of defective units in the lots which have been treated by this sampling plan never exceeds $AOQL$. From the consumer's point of view, therefore, it is desirable to specify that the parameters should be chosen in such a way that $AOQL$ has a certain fixed low value.

Finally, it is important also to consider the cost of carrying out the work of testing. This makes it desirable to choose the parameters for the plan in such a way that the number of tested units per lot is kept low. Since the testing of the sample units may stop as soon as $n-r$ satisfactory units have been found, the mean value of the number of tested units will be at most equal to

$$n f(p) + N(1 - f(p)),$$

where $f(p)$ is given by 16.6.1. Thus one might impose the requirement on the parameters that this function (which is called the plan's ASN, or *average sampling number*) shall be as small as possible for a certain given p.

In the plan which has been discussed above, one has only the two parameters n and r at his disposal, and therefore one must be content with determining them so that two given conditions are satisfied. By introducing more complicated plans, in which one takes two or more samples from the same lot, according to certain rules, the number of parameters can be increased and thereby the flexibility of the plan. In this connection, the *sequential sampling* developed by WALD is highly important. In this method, the sampling is carried out step by step with one unit at a time, and after each step one decides by certain rules, on the basis of the tests already made, whether to accept the lot, reject it, or continue the sampling. The method has interesting and practically useful properties. Among other things, it gives an ASN which is only about half as large as the corresponding value for a plan of the kind described above, and of efficiency comparable in other respects.

The theory of these more refined sampling methods is, however, too complicated to be treated here. We refer the reader to the works of DODGE-ROMIG and WALD mentioned in the bibliography. In the first-named work, one finds also detailed tables for determining parameters in the simple testing scheme treated here, starting with various requirements of the kind discussed here.

16.7. Problems. — 1. — In a set of chemical experiments designed to test the composition of certain raw materials used in textile industry, the following figures were obtained (data supplied by Mr. G. BLOM). x is a cholorimetric measure, while y measures the quantity of bichromate consumed in a certain chemical reaction. It is required to study the regression of y on x.

x	y
24	4.93
29	5.98
36	7.40
45	8.88
52	10.36
58	11.96
61	12.32
72	14.53
78	15.78
86	17.26
90	17.94
97	19.23
103	20.71
119	23.92

2. In order to determine the unknown angles α_1, α_2, and α_3 of a certain triangle, measurements have been made on each angle, yielding the observed values x_1, x_2, and x_3, of weights p_1, p_2, and p_3 respectively. Use 16.2 to find the "best" estimates of the angles, and their standard errors.

3. Six machines used for wool combing were tested for the percentage of noils. During each of eleven successive weeks, one test was performed with each machine, the following data being obtained (data supplied by Mr. G. BLOM). Apply analysis of variance to test whether there are significant differences between a) machines, b) weeks.

Week	Machine					
	A	B	C	D	E	F
1	10.6	11.4	10.4	12.2	14.8	12.4
2	9.2	9.6	7.8	10.0	11.8	9.6
3	7.6	8.4	8.0	9.4	9.4	9.2
4	7.6	8.6	8.4	9.6	12.0	10.0
5	7.6	8.8	9.4	9.2	11.4	10.4
6	9.8	11.8	11.0	12.2	14.6	14.8
7	6.0	7.6	6.6	7.4	9.0	9.0
8	10.6	15.0	11.0	15.0	14.6	18.0
9	9.0	11.0	9.0	12.0	12.0	14.0
10	6.6	9.6	8.6	13.0	12.6	13.0
11	8.4	10.4	8.0	11.6	10.6	16.0

4. In the study of stratified sampling made in 16.5, only the case of representative sampling has been considered. In this case, the number n_i of individuals drawn from the i-th stratum is $n_i = n\,p_i$.

Another possible way of determining the numbers n_i would be to lay down the condition that the variance of the estimate $\bar{\bar{x}}$ as given by 16.5.4:

$$D^2(\bar{\bar{x}}) = \sum_{i=1}^{r} p_i^2 \frac{N_i - n_i}{N_i - 1} \cdot \frac{\sigma_i^2}{n_i},$$

should be rendered as small as possible. Work out the solution for this case, and comment on the result.

5. From a batch of N units, containing $100\ p$ % defectives, a sample of n units is drawn at random, where n is small compared to N. The batch is accepted if the sample contains at most one defective unit. How large should n be taken in order that the probability of accepting a batch with 10 % defectives should be smaller than 5 % ? What is, for the smallest n satisfying this condition, the probability of accepting a batch containing 1 % defectives ?

APPENDIX

The main formulas connected with the Gamma function which have been used in this book, particularly in chapters 6–8, will be briefly discussed in this Appendix. Complete proofs will be found in most treatises on Integral Calculus, and also in *Mathematical Methods*, chapter 12.

The Gamma function $\Gamma(p)$ is defined for all $p > 0$ by the integral

$$(1) \qquad \Gamma(p) = \int_0^\infty x^{p-1} e^{-x} \, dx.$$

The function takes positive values for all $p > 0$, and tends to infinity as p tends to zero or to $+\infty$. There is a unique minimum at $p = 1.4616$, where the function takes the value 0.8856. By means of partial integration the formula

$$\Gamma(p+1) = p\,\Gamma(p)$$

is obtained. For integral values of p, this gives

$$\Gamma(p+1) = p!$$

Substituting in (1) ax for x, where $a > 0$, we obtain

$$(2) \qquad \int_0^\infty x^{p-1} e^{-\alpha x} \, dx = \frac{\Gamma(p)}{\alpha^p}.$$

A further important integral formula is

$$\int_0^1 x^{p-1} (1-x)^{q-1} \, dx = \frac{\Gamma(p)\,\Gamma(q)}{\Gamma(p+q)},$$

where p and q are both positive. Taking here $p = q = \frac{1}{2}$, and introducing the new variable $y = 2x - 1$, we obtain

$$\Gamma^2(\tfrac{1}{2}) = \int_0^1 \frac{dx}{\sqrt{x(1-x)}} = \int_{-1}^1 \frac{dy}{\sqrt{1-y^2}} = \pi,$$

$$\Gamma(\tfrac{1}{2}) = \sqrt{\pi}.$$

For $p = \frac{1}{2}$ and $x = y^2$ we thus obtain from (2)

$$(3) \qquad \int_{-\infty}^{\infty} e^{-\alpha y^2}\, dy = \sqrt{\frac{\pi}{\alpha}}.$$

For large values of p, there is an important asymptotic formula for the Gamma function, which is known as STIRLING's formula. Taking $p = n + 1$, where n is an integer, this formula can be written

$$\log n! = (n + \tfrac{1}{2}) \log n - n + \tfrac{1}{2} \log 2\pi + \frac{\theta}{12\,n},$$

or

$$(4) \qquad n! = \left(\frac{n}{e}\right)^n \sqrt{2\pi n}\; e^{\frac{\theta}{12\,n}},$$

where θ lies between 0 and 1.

REFERENCES

for further reading.

ARLEY, N. and BUCH, K. R., Introduction to the theory of probability and statistics. New York (Wiley) 1950.

COCHRAN, W. G. and COX, G. M., Experimental designs. New York (Wiley) 1950.

CRAMÉR, H., Random variables and probability distributions. Cambridge University Press 1937.

——, Mathematical methods of statistics. Princeton University Press 1946.

DODGE, H. and ROMIG, H., Sampling inspection tables. New York (Wiley) 1944.

DOOB, J. L., Stochastic processes. New York (Wiley) 1953.

ELDERTON, W. P., Frequency curves and correlation. Cambridge University Press 1938 (third ed.).

FELLER, W., An introduction to probability theory and its applications. Vol. I, New York (Wiley) 1950.

FISHER, R. A., Statistical methods for research workers. Edinburgh (Oliver and Boyd) 1950 (eleventh ed.).

——, The design of experiments. Edinburgh (Oliver and Boyd) 1949 (fifth ed.).

—— and YATES, F., Statistical tables. London 1943 (second ed.).

HALD, A., Statistical theory with engineering applications. New York (Wiley) 1952.

HANSEN, M. H., HURWITZ, W. N. and MADOW, W. G., Sample survey methods and theory, I–II. New York (Wiley) 1953.

HOEL, P. G., Mathematical statistics. New York (Wiley) 1947.

KEMPTHORNE, O., The design and analysis of experiments. New York (Wiley) 1952.

KENDALL, M. G., The advanced theory of statistics, I–II. London (Griffin) 1945–1948.

KOLMOGROFF, A., Grundbegriffe der Wahrscheinlichkeitsrechnung. Berlin (Springer) 1933.

LÉVY, P., Théorie de l'addition des variables aléatoires. Paris (Gauthier-Villars) 1937.

MISES, R. v., Wahrscheinlichkeitsrechnung. Wien (Deuticke) 1931.

——, Wahrscheinlichkeit, Statistik und Wahrheit. Wien (Springer) 1951 (third ed.). English translation: Probability, statistics, truth. London 1939.

NEYMAN, J., First course in probability and statistics. New York (Holt) 1950.

TODHUNTER, I., A history of the mathematical theory of probability. Cambridge (Macmillan) 1865.

WALD, A.. Sequential analysis. New York (Wiley) 1947.

——, Statistical decision functions. New York (Wiley) 1950.

WILKS, S. S., Mathematical statistics. Princeton University Press 1944.

265

ANSWERS TO PROBLEMS

3.7. — 1. a) $\dfrac{1}{5}$, b) $\dfrac{2}{9}$.

2. $\dfrac{13}{36}$.

3. $\dfrac{15}{34}$.

4. $\dfrac{1}{2}\left(\dfrac{\nu_1}{n_1}+\dfrac{\nu_2}{n_2}\right)$.

5. a) $1-P(AB)$, b) $1-P(A)-P(B)+P(AB)$, c) $1-P(A)+P(AB)$,
   d) $P(B)-P(AB)$, e) $=b)$, f) $=a)$, g) $=d)$, h) $P(A)+P(B)-P(AB)$.

6. $p_0=1-S_1+S_2$, $p_1=S_1-2S_2$, $p_2=S_2$, where $S_1=P(A)+P(B)$,
   $S_2=P(AB)$.

7. $p_0=1-S_1+S_2-S_3$, $p_1=S_1-2S_2+3S_3$, $p_2=S_2-3S_3$, $p_3=S_3$, where
   $S_1=P(A)+P(B)+P(C)$, $S_2=P(AB)+P(AC)+P(BC)$, $S_3=P(ABC)$.

8. 0.504, 0.398, 0.092, 0.006.

13. $P(A)=P(B)=P(C)=\frac{1}{2}$, $P(AB)=P(AC)=P(BC)=\frac{1}{4}$, $P(ABC)=\frac{1}{4}$.

14. $\frac{1}{2}$.

4.4. — 1. $\dfrac{5}{16}$.

2. a) 0.862, b) 0.824.

3. a) $\dfrac{27}{64}$, b) $\dfrac{9139}{20825}$.

4. $\dfrac{175}{256}$.

5. $p_0=p_1=\left(\dfrac{5}{6}\right)^5$.

6. $\dfrac{7}{64}$.

7. $\dbinom{r}{k} \dfrac{(n-1)^{r-k}}{n^r}$.

8. a) $\dfrac{43}{216}$, b) $\dfrac{20}{216} = \dfrac{5}{54}$.

9. $\dfrac{5}{16}$.

10. 0.09666.

11. 69.

12. $\dfrac{25}{57}, \dfrac{26}{57}, \dfrac{6}{57}$.

13. a) $\dfrac{55}{575}$, b) $\dfrac{286}{575}$, c) $\dfrac{234}{575}$.

14. 0.002188.

15. $\dfrac{98}{9889}$.

16. $\dfrac{\nu}{n} \cdot \dbinom{n}{\nu} p^\nu q^{n-\nu}$.

17. a) $\left(\dfrac{k}{6}\right)^n - \left(\dfrac{k-1}{6}\right)^n$, b) $\left(1 - \dfrac{k-1}{6}\right)^n - \left(1 - \dfrac{k}{6}\right)^n$.

18. $\dfrac{21}{32}$.

19. $\dfrac{2}{27}$.

20. $\dfrac{10!}{10^{10}} = 0.00036$.

21. $1 - \dbinom{k}{1}\left(\dfrac{k-1}{k}\right)^n + \dbinom{k}{2}\left(\dfrac{k-2}{k}\right)^n - \cdots + (-1)^{k-1}\dbinom{k}{k-1}\left(\dfrac{1}{k}\right)^n$.

22. $\dfrac{1}{2}\left(1 + \left(1 - \dfrac{2}{N}\right)^n\right)$.

23. $p_n = r_n = \frac{1}{6}\left(1-\left(-\frac{1}{2}\right)^n\right), \quad q_n = \frac{1}{3}\left(2+\left(-\frac{1}{2}\right)^n\right).$

24. $P_n = \frac{1}{2}+(2p-1)^n\left(P_0-\frac{1}{2}\right).$

25. a) $\dfrac{i^\nu(N-i)^{n-\nu}}{\displaystyle\sum_{i=0}^{N} i^\nu(N-i)^{n-\nu}}$, \quad b) $\dfrac{\displaystyle\sum_{i=0}^{N} i^{\nu+1}(N-i)^{n-\nu}}{N\displaystyle\sum_{i=0}^{N} i^\nu(N-i)^{n-\nu}} \to \dfrac{\nu+1}{n+2}.$

5.11. — 1. For $x>0$ we have $P(Y\leqq x)=F\left(\sqrt{x}\right)-F\left(-\sqrt{x}\right),$

$P(Z\leqq x)=F(\log x).$

3. CAUCHY's distribution.

4. Fr. f.: $\dfrac{\nu}{(b-a)^n}\dbinom{n}{\nu}(x-a)^{\nu-1}(b-x)^{n-\nu},$ \quad Mean: $a+\dfrac{\nu}{n+1}(b-a).$

6. $c=$ the median.

7. $E(n)=\dfrac{1}{p}, \quad D^2(n)=\dfrac{q}{p^2}.$

8. Cf. 16.5.

10. 1, 1.

11. Frequency functions: a) $\log\dfrac{1}{x}, \quad 0<x<1,$ \quad b) $|1-x|, \quad -1<x<1$

c) $2(1-x), \quad 0<x<1.$

12. $\dfrac{5}{9}.$

13. $\dfrac{5+6\log 2}{36}.$

14. $\dfrac{4r}{\pi}.$

15. $\dfrac{4r}{\pi}.$

16. a) 1 for $k<\frac{1}{3}$, $1-(3k-1)^2$ for $\frac{1}{3}<k<\frac{1}{2}$, $3(1-k)^2$ for $k>\frac{1}{2}$,

b) 0 for $k<\frac{1}{3}$, $6(3k-1)$ for $\frac{1}{3}<k<\frac{1}{2}$, $6(1-k)$ for $k>\frac{1}{2}$.

19. $\dfrac{p^\nu e^{\nu t}(1-pe^t)}{1-e^t+p^\nu q e^{(\nu+1)t}}, \quad E(n)=\dfrac{1-p^\nu}{p^\nu q}.$

268

20. $\dfrac{(k-1)!\, e^{kt}}{\displaystyle\prod_{\nu=1}^{k-1}(k-\nu\, e^{t})}$, $E(n)=k\left(1+\dfrac{1}{2}+\cdots+\dfrac{1}{k}\right)\cdot$

6.6. — 2.

k	Exact	Normal	Tchebycheff
1	0.32925	0.31731	1
2	0.06391	0.04550	0.25
3	0.00154	0.00270	0.1111
4	0.00004	0.00006	0.0625
5	0.00000	0.00000	0.04

5. $p_n(t)=\dbinom{n-1}{n_0-1}e^{-\lambda n_0 t}(1-e^{-\lambda t})^{n-n_0}.$

7.8. — 1. $m=74.35\ \%$, $\sigma=3.22\ \%.$

2. $\sigma\sqrt{\dfrac{2}{\pi}}\cdot$

4. $\dfrac{\sigma}{\sqrt{\pi}}\cdot$

5. a) $\tfrac{1}{2}n$, b) $\lambda_1\sqrt{\dfrac{n}{12}}=2.5758\sqrt{\dfrac{n}{12}}\cdot$

6. $a=a_1-\dfrac{\sqrt{\alpha_2-\alpha_1^2}}{\eta}$, $\sigma^2=\log(1+\eta^2)$,

   $m=\log(\alpha_1-a)-\tfrac{1}{2}\sigma^2.$

7. $a=\dfrac{\zeta_1\zeta_3-\zeta_2^2}{\zeta_1+\zeta_3-2\zeta_2}$, $m=\log\dfrac{(\zeta_3-\zeta_2)(\zeta_2-\zeta_1)}{\zeta_1+\zeta_3-2\zeta_2}$, $\sigma=\dfrac{1}{\lambda_{50}}\log\dfrac{\zeta_3-\zeta_2}{\zeta_2-\zeta_1}$,

   where $\lambda_{50}=0.6745$ is the 50 % value of the normal distribution.

8.5. — 1. For χ^2: $\gamma_1=\sqrt{\dfrac{8}{n}}$, $\gamma_2=\dfrac{12}{n}\cdot$ For t: $\gamma_1=0$, $\gamma_2=\dfrac{6}{n-4}\cdot$

2. $E(F)=\dfrac{n}{n-2}$, $D^2(F)=\dfrac{2n^2(m+n-2)}{m(n-2)^2(n-4)}\cdot$

4. t and v have the Student and F distributions respectively.

   u has the fr. f. $C\left(1-\dfrac{x^2}{m}\right)^{\frac{m-3}{2}}$, $-\sqrt{m}<x<\sqrt{m}.$

   w has the fr. f. $Cx^{\frac{m}{2}-1}(1-x)^{\frac{n}{2}-1}$, $0<x<1.$

269

9.5. — 2.

		m_1	m_2	σ_1^2	σ_2^2	ϱ
a)	0	$\dfrac{1}{3}$	$\dfrac{1}{3}$	$\dfrac{4}{45}$		0
b)	0	0	$\dfrac{1}{3}$	$\dfrac{1}{7}$	$\dfrac{\sqrt{21}}{5}$	
c)	0	0	$\dfrac{1}{3}$	$\dfrac{1}{2}$	$\dfrac{4\sqrt{6}}{\pi^2}$	
d)	0	$\dfrac{2}{\pi}$	$\dfrac{1}{2}$	$\dfrac{1}{2}-\dfrac{4}{\pi^2}$		0

3. $\operatorname{tg} v = \dfrac{\sigma_1\sigma_2}{\sigma_1^2+\sigma_2^2}\cdot\dfrac{1-\varrho^2}{\varrho}.$

4. $a c\,\sigma_1^2 + (a d + b c)\,\varrho\,\sigma_1\sigma_2 + b d\,\sigma_2^2 = 0.$

5. $\left[\left(1+\dfrac{s^2}{p q\,(m_1-m_2)^2}\right)\left(1+\dfrac{s^2}{p q\,(n_1-n_2)^2}\right)\right]^{-\frac{1}{2}}.$

6. a) $\dfrac{1}{\sqrt{n}},$ b) 0.

7. $m_1=3,\;\; m_2=\tfrac{1}{2},\;\; \sigma_1^2=3,\;\; \sigma_2^2=\tfrac{3}{4},\;\; \varrho=-\tfrac{1}{3},\;\; E(Y\,|\,X=x)=\dfrac{1}{x},$

   $E(X\,|\,Y=y)=\dfrac{4}{y+1},\;\; \eta_{YX}^2=\tfrac{1}{3},\;\; \eta_{XY}^2=\tfrac{1}{5}.$

8. $m_1=4,\;\; m_2=7,\;\; \varrho=-\tfrac{1}{2}.$

9. $e^{-c^2}.$

11. The conditional distributions are rectangular, the regression curves coincide with the least squares regression lines, and the correlation coefficient is ϱ.

10.5. — 1. $\varrho_{ik}=-\dfrac{1}{3},\;\; \varrho_{ij\cdot k}=-\dfrac{1}{2},\;\; \varrho_{ij\cdot kl}=-1,\;\; \varrho_{i(jk)}=\dfrac{1}{\sqrt{3}},\;\; \varrho_{i(jkl)}=1.$

2. $\varrho_{12}=\varrho_{23}=-\dfrac{1}{3},\;\; \varrho_{13}=\dfrac{1}{9},\;\; \varrho_{12\cdot3}=\varrho_{23\cdot1}=-\dfrac{1}{\sqrt{10}},$

   $\varrho_{13\cdot2}=0,\;\; \varrho_{1(23)}=\varrho_{3(12)}=\dfrac{1}{3},\;\; \varrho_{2(13)}=\dfrac{1}{\sqrt{5}}.$

3. $-\tfrac{1}{2}\leqq c\leqq 1.$

4. $\varrho_{12}=-\dfrac{a^2+b^2-c^2}{2\,a b}.$

5. $\varrho_{12}+\varrho_{13}+\varrho_{23}=-1.$

13.5. — 2. The fr. f. of u is $C \left(1 - \dfrac{x^2}{n-1}\right)^{\frac{n-4}{2}}$, $\quad -\sqrt{n-1} < x < \sqrt{n-1}$.

3. $v \sqrt{\dfrac{n_1 n_2 (n_1 + n_2 - 2)}{b^2 n_1 + a^2 n_2}}$ has Student's distribution with $n_1 + n_2 - 2$ degrees of freedom.

4. $\dfrac{n_1 (n_2 - 1)}{n_2 (n_1 - 1)} w$ has the F distribution with $n_1 - 1$ degrees of freedom in the numerator and $n_2 - 1$ in the denominator.

5. $n \begin{pmatrix} n-1 \\ n-i \end{pmatrix} [F(x)]^{i-1} [1 - F(x)]^{n-i} f(x)$.

6. $E(x_i) = \dfrac{i}{n+1}$, $\quad D^2(x_i) = \dfrac{i(n-i+1)}{(n+1)^2 (n+2)}$.

14.7. — 2. If the distribution ranges over the interval $(a-h,\ a+h)$, we have
$$D^2(\bar{x}) = \frac{h^2}{3n}, \quad D^2(z_2) = \frac{h^2}{n+2}, \quad D^2\left(\frac{x_{max} + x_{min}}{2}\right) = \frac{2 h^2}{(n+1)(n+2)}.$$

3. The confidence limits are 126.9786 and 126.9849 (from 14.4.3).

4. 0.058 and 0.142.

5. $\bar{x} \pm \lambda_p \dfrac{\bar{x}}{\sqrt{n}}$.

6. 22.08 and 55.12.

7. 0.0098 and 0.0266.

8. $\dfrac{1 - e^{-\lambda}}{\lambda} = \dfrac{n - f_0}{n \bar{x}}$, $\quad p = 1 - \dfrac{\bar{x}}{\lambda}$.

9. The interval (16, 26) has the confidence coefficient 95.1 %.

10. $\dfrac{r}{n} \pm \lambda_p \sqrt{\dfrac{r(n-r)}{n^3}}$, $\quad$ where $x_r \leq x$, $x_{r+1} > x$.

15.3. — 1. $\chi^2 = 4.087$ with 3 d. of fr.

2. $q = 0.08706$, $\chi^2 = 3.087$ with 2 d. of fr.

4. $\lambda = 1.0374$, $p = 0.61489$. Pooling the groups 4–6, we obtain $\chi^2 = 1.878$ with 2 d. of fr.

5. a) $\chi^2 = 3.2$ with 1 d. of fr. b) $\sqrt{2\chi^2} - \sqrt{359} = -1.471$.

271

16.7. — 2. $\alpha_1 = \dfrac{p_2 \, p_3 \, (\pi - x_1 - x_2 - x_3) \pm \sqrt{p_1 \, p_2 \, p_3 \, (p_2 + p_3)} \, | \pi - x_1 - x_2 - x_3 |}{p_1 \, p_2 + p_1 \, p_3 + p_2 \, p_3}$.

3. Mean squares: between machines 28.90, between weeks 20.10. error 1.27.

4. Minimum of $D^2 (\bar{\bar{x}})$ is obtained for $n_i = k \, p_i \, \sigma_i \sqrt{\dfrac{N_i}{N_i - 1}}$ where k is determined by the condition $\sum_1^r n_i = n$.

5. a) $n = 46$,   b) 0.923.

TABLE I
THE NORMAL DISTRIBUTION

$$\Phi(x) = \frac{1}{\sqrt{2\pi}} \int_{-\infty}^{x} e^{-\frac{t^2}{2}} dt, \quad \varphi(x) = \Phi'(x) = \frac{1}{\sqrt{2\pi}} e^{-\frac{x^2}{2}}.$$

For $x < 0$, the functions are found by means of the relations

$$\Phi(-x) = 1 - \Phi(x), \quad \varphi(-x) = \varphi(x), \quad \varphi^{(i)}(-x) = (-1)^i \varphi^{(i)}(x).$$

x	$\Phi(x)$	$\varphi(x)$	$\varphi'(x)$	$\varphi''(x)$	$\varphi^{(3)}(x)$	$\varphi^{(4)}(x)$	$\varphi^{(5)}(x)$	$\varphi^{(6)}(x)$
0.0	0.50000	0.39894	-0.00000	-0.39894	$+0.00000$	$+1.19683$	-0.00000	-5.98413
0.1	0.53983	0.39695	0.03970	0.39298	0.11869	1.16708	0.59146	5.77625
0.2	0.57926	0.39104	0.07821	0.37540	0.23150	1.07990	1.14197	5.17112
0.3	0.61791	0.38139	0.11442	0.34706	0.33295	0.94130	1.61420	4.22226
0.4	0.65542	0.36827	0.14731	0.30935	0.41835	0.76070	1.97770	3.01241
0.5	0.69146	0.35207	0.17603	0.26405	0.48409	0.55010	2.21141	1.64481
0.6	0.72575	0.33322	0.19993	0.21326	0.52783	0.32309	2.30517	-0.23237
0.7	0.75804	0.31225	0.21858	0.15925	0.54863	$+0.09371$	2.26012	$+1.11354$
0.8	0.78814	0.28969	0.23175	0.10429	0.54694	-0.12468	2.08800	2.29382
0.9	0.81594	0.26609	0.23948	-0.05056	0.52445	0.32034	1.80951	3.23026
1.0	0.84134	0.24197	0.24197	0.00000	0.48394	0.48394	1.45182	3.87153
1.1	0.86433	0.21785	0.23964	$+0.04575$	0.42895	0.60909	1.04580	4.19585
1.2	0.88493	0.19419	0.23302	0.08544	0.36352	0.69255	0.62301	4.21034
1.3	0.90320	0.17137	0.22278	0.11824	0.29184	0.73413	-0.21300	3.94753
1.4	0.91924	0.14973	0.20962	0.14374	0.21800	0.73642	$+0.15897$	3.45953
1.5	0.93319	0.12952	0.19428	0.16190	0.14571	0.70425	0.47355	2.81094
1.6	0.94520	0.11092	0.17747	0.17304	0.07809	0.64405	0.71813	2.07125
1.7	0.95543	0.09405	0.15988	0 17775	$+0.01759$	0.56316	0.88702	1.30785
1.8	0.96407	0.07895	0.14211	0.17685	-0.03411	0.46915	0.98090	$+0.58014$
1.9	0.97128	0.06562	0.12467	0.17126	0.07605	0.36928	1.00583	-0.06467
2.0	0.97725	0.05399	0.10798	0.16197	0.10798	0.26996	0.97184	0.59390
2.1	0.98214	0.04398	0.09237	0.14998	0.13024	0.17646	0.89150	0.98987
2.2	0.98610	0.03547	0.07804	0.13622	0.14360	0.09274	0.77844	1.24885
2.3	0.98928	0.02833	0.06515	0.12152	0.14920	-0.02141	0.64604	1.37883
2.4	0.99180	0.02239	0.05375	0.10660	0.14834	$+0.03623$	0.50642	1.39654
2.5	0.99379	0.01753	0.04382	0.09202	0.14242	0.07997	0.36974	1.32421
2.6	0.99534	0.01358	0.03532	0.07824	0.13279	0.11053	0.24376	1.18645
2.7	0.99653	0.01042	0.02814	0.06555	0.12071	0.12926	0.13381	1.00761
2.8	0.99744	0.00792	0.02216	0.05414	0.10727	0.13793	$+0.04287$	-0.80970
2.9	0.99813	0.00595	0.01726	0.04411	0.09339	0.13850	-0.02810	$+0.61102$
3.0	0.99865	0.00443	0.01330	0.03545	0.07977	0.13296	0.07977	0.42546
3.1	0.99903	0.00327	0.01013	0.02813	0.06694	0.12313	0.11395	0.26242
3.2	0.99931	0.00238	0.00763	0.02203	0.05523	0.11066	0.13319	0.12712
3.3	0.99952	0.00172	0.00568	0.01704	0.04485	0.09690	0.14036	0.02130
3.4	0.99966	0.00123	0.00419	0.01301	0.03586	0.08290	0.13840	0.05607
3.5	0.99977	0.00087	0.00305	0.00982	0.02825	0.06943	0.13000	0.10784
3.6	0.99984	0.00061	0.00220	0.00732	0.02194	0.05703	0.11755	0.13802
3.7	0.99989	0.00042	0.00157	0.00539	0.01680	0.04599	0.10297	0.15102
3.8	0.99993	0.00029	0.00111	0.00392	0.01269	0.03646	0.08777	0.15124
3.9	0.99995	0.00020	0.00077	0.00282	0.00946	0.02842	0.07302	0.14264
4.0	0.99997	0.00013	-0.00054	$+0.00201$	-0.00696	$+0.02181$	-0.05942	$+0.12861$

TABLE II

THE NORMAL DISTRIBUTION

For a given percentage p, the p % value λ_p of the normal distribution is defined by the condition (cf. 7.2)

$$P\left(\left|X-m\right|>\lambda_p\,\sigma\right)=\frac{p}{100},$$

where X is normally distributed, with mean m and standard deviation σ. Thus the probability that X differs from its mean in either direction by more than λ_p times the s.d. is equal to p %.

λ_p as function of p		p as function of λ_p	
p	λ_p	λ_p	p
100	0.0000	0.0	100.000
95	0.0627	0.2	84.148
90	0.1257	0.4	68.916
85	0.1891	0.6	54.851
80	0.2533	0.8	42.371
75	0.3186	1.0	31.731
70	0.3853	1.2	23.014
65	0.4538	1.4	16.151
60	0.5244	1.6	10.960
55	0.5978	1.8	7.186
50	0.6745	2.0	4.550
45	0.7554	2.2	2.781
40	0.8416	2.4	1.640
35	0.9346	2.6	0.932
30	1.0364	2.8	0.511
25	1.1503	3.0	0.270
20	1.2816	3.2	0.137
15	1.4395	3.4	0.067
10	1.6449	3.6	0.032
5	1.9600	3.8	0.014
1	2.5758	4.0	0.006
0.1	3.2905		
0.01	3.8906		

TABLE III

THE χ^2 DISTRIBUTION

For a given percentage p, the $p\%$ value χ_p^2 of the χ^2 distribution is defined by the condition (cf. 8.1)

$$P(\chi^2 > \chi_p^2) = \frac{p}{100}.$$

Thus the probability that χ^2 assumes a value exceeding χ_p^2 is equal to $p\%$.

De-grees of free-dom n	χ_p^2 as function of n and p													
	$p=99$	98	95	90	80	70	50	30	20	10	5	2	1	0.1
1	0.000	0.001	0.004	0.016	0.064	0.148	0.455	1.074	1.642	2.706	3.841	5.412	6.635	10.827
2	0.020	0.040	0.103	0.211	0.446	0.713	1.386	2.408	3.219	4.605	5.991	7.824	9.210	13.815
3	0.115	0.185	0.352	0.584	1.005	1.424	2.366	3.665	4.642	6.251	7.815	9.837	11.341	16.268
4	0.297	0.429	0.711	1.064	1.649	2.195	3.357	4.878	5.989	7.779	9.488	11.668	13.277	18.465
5	0.554	0.752	1.145	1.610	2.343	3.000	4.351	6.064	7.289	9.236	11.070	13.388	15.086	20.517
6	0.872	1.134	1.635	2.204	3.070	3.828	5.348	7.231	8.558	10.645	12.592	15.033	16.812	22.457
7	1.239	1.564	2.167	2.833	3.822	4.671	6.346	8.383	9.803	12.017	14.067	16.622	18.475	24.322
8	1.646	2.032	2.733	3.490	4.594	5.527	7.344	9.524	11.030	13.362	15.507	18.168	20.090	26.125
9	2.088	2.532	3.325	4.168	5.380	6.393	8.343	10.656	12.242	14.684	16.919	19.679	21.666	27.877
10	2.558	3.059	3.940	4.865	6.179	7.267	9.342	11.781	13.442	15.987	18.307	21.161	23.209	29.588
11	3.053	3.609	4.575	5.578	6.989	8.148	10.341	12.899	14.631	17.275	19.675	22.618	24.725	31.264
12	3.571	4.178	5.226	6.304	7.807	9.034	11.340	14.011	15.812	18.549	21.026	24.054	26.217	32.909
13	4.107	4.765	5.892	7.042	8.634	9.926	12.340	15.119	16.985	19.812	22.362	25.472	27.688	34.528
14	4.660	5.368	6.571	7.790	9.467	10.821	13.339	16.222	18.151	21.064	23.685	26.873	29.141	36.123
15	5.229	5.985	7.261	8.547	10.307	11.721	14.339	17.322	19.311	22.307	24.996	28.259	30.578	37.697
16	5.812	6.614	7.962	9.312	11.152	12.624	15.338	18.418	20.465	23.542	26.296	29.633	32.000	39.252
17	6.408	7.255	8.672	10.085	12.002	13.531	16.338	19.511	21.615	24.769	27.587	30.995	33.409	40.790
18	7.015	7.906	9.390	10.865	12.857	14.440	17.338	20.601	22.760	25.989	28.869	32.346	34.805	42.312
19	7.633	8.567	10.117	11.651	13.716	15.352	18.338	21.689	23.900	27.204	30.144	33.687	36.191	43.820
20	8.260	9.237	10.851	12.443	14.578	16.266	19.337	22.775	25.038	28.412	31.410	35.020	37.566	45.315
21	8.897	9.915	11.591	13.240	15.445	17.182	20.337	23.858	26.171	29.615	32.671	36.343	38.932	46.797
22	9.542	10.600	12.338	14.041	16.314	18.101	21.337	24.939	27.301	30.813	33.924	37.659	40.289	48.268
23	10.196	11.293	13.091	14.848	17.187	19.021	22.337	26.018	28.429	32.007	35.172	38.968	41.638	49.728
24	10.856	11.992	13.848	15.659	18.062	19.943	23.337	27.096	29.553	33.196	36.415	40.270	42.980	51.179
25	11.524	12.697	14.611	16.473	18.940	20.867	24.337	28.172	30.675	34.382	37.652	41.566	44.314	52.620
26	12.198	13.409	15.379	17.292	19.820	21.792	25.336	29.246	31.795	35.563	38.885	42.856	45.642	54.052
27	12.879	14.125	16.151	18.114	20.703	22.719	26.336	30.319	32.912	36.741	40.113	44.140	46.963	55.476
28	13.565	14.847	16.928	18.939	21.588	23.647	27.336	31.391	34.027	37.916	41.337	45.419	48.278	56.893
29	14.256	15.574	17.708	19.768	22.475	24.577	28.336	32.461	35.139	39.087	42.557	46.693	49.588	58.302
30	14.953	16.306	18.493	20.599	23.364	25.508	29.336	33.530	36.250	40.256	43.773	47.962	50.892	59.703

TABLE IV

THE t DISTRIBUTION

For **a** given percentage p, the p % value t_p of the t distribution is defined by the condition (cf. 8.2)

$$P(|t| > t_p) = \frac{p}{100}.$$

Thus the probability that t differs from its mean zero in either direction by more than t_p is equal to p %.

Degrees of freedom n	t_p as function of n and p												
	$p = 90$	80	70	60	50	40	30	20	10	5	2	1	0.1
1	0.158	0.325	0.510	0.727	1.000	1.376	1.963	3.073	6.314	12.706	31.821	63.657	636.619
2	0.142	0.289	0.445	0.617	0.816	1.061	1.386	1.886	2.920	4.303	6.965	9.925	31.589
3	0.137	0.277	0.424	0.584	0.765	0.978	1.250	1.638	2.353	3.182	4.541	5.841	12.941
4	0.134	0.271	0.414	0.569	0.741	0.941	1.190	1.533	2.132	2.776	3.747	4.604	8.610
5	0.132	0.267	0.408	0.559	0.727	0.920	1.156	1.476	2.015	2.571	3.365	4.032	6.859
6	0.131	0.265	0.404	0.553	0.718	0.906	1.134	1.440	1.943	2.447	3.143	3.707	5.959
7	0.130	0.263	0.402	0.549	0.711	0.896	1.119	1.415	1.895	2.365	2.998	3.499	5.405
8	0.130	0.262	0.399	0.546	0.706	0.889	1.108	1.397	1.860	2.306	2.896	3.355	5.041
9	0.129	0.261	0.398	0.543	0.703	0.883	1.100	1.383	1.833	2.262	2.821	3.250	4.781
10	0.129	0.260	0.397	0.542	0.700	0.879	1.093	1.372	1.812	2.228	2.764	3.169	4.587
11	0.129	0.260	0.396	0.540	0.697	0.876	1.088	1.363	1.796	2.201	2.718	3.106	4.437
12	0.128	0.259	0.395	0.539	0.695	0.873	1.083	1.356	1.782	2.179	2.681	3.055	4.318
13	0.128	0.259	0.394	0.538	0.694	0.870	1.079	1.350	1.771	2.160	2.650	3.012	4.221
14	0.128	0.258	0.393	0.537	0.692	0.868	1.076	1.345	1.761	2.145	2.624	2.977	4.140
15	0.128	0.258	0.393	0.536	0.691	0.866	1.074	1.341	1.753	2.131	2.602	2.947	4.073
16	0.128	0.258	0.392	0.535	0.690	0.865	1.071	1.337	1.746	2.120	2.583	2.921	4.015
17	0.128	0.257	0.392	0.534	0.689	0.863	1.069	1.333	1.740	2.110	2.567	2.898	3.965
18	0.127	0.257	0.392	0.534	0.688	0.862	1.067	1.330	1.734	2.101	2.552	2.878	3.922
19	0.127	0.257	0.391	0.533	0.688	0.861	1.066	1.328	1.729	2.093	2.539	2.861	3.883
20	0.127	0.257	0.391	0.533	0.687	0.860	1.064	1.325	1.725	2.086	2.528	2.845	3.850
21	0.127	0.257	0.391	0.532	0.686	0.859	1.063	1.323	1.721	2.080	2.518	2.831	3.819
22	0.127	0.256	0.390	0.532	0.686	0.858	1.061	1.321	1.717	2.074	2.508	2.819	3.792
23	0.127	0.256	0.390	0.532	0.685	0.858	1.060	1.319	1.714	2.069	2.500	2.807	3.767
24	0.127	0.256	0.390	0.531	0.685	0.857	1.059	1.318	1.711	2.064	2.492	2.797	3.745
25	0.127	0.256	0.390	0.531	0.684	0.856	1.058	1.316	1.708	2.060	2.485	2.787	3.725
26	0.127	0.256	0.390	0.531	0.684	0.856	1.058	1.315	1.706	2.056	2.479	2.779	3.707
27	0.127	0.256	0.389	0.531	0.684	0.855	1.057	1.314	1.703	2.052	2.473	2.771	3.690
28	0.127	0.256	0.389	0.530	0.683	0.855	1.056	1.313	1.701	2.048	2.467	2.763	3.674
29	0.127	0.256	0.389	0.530	0.683	0.854	1.055	1.311	1.699	2.045	2.462	2.756	3.659
30	0.127	0.256	0.389	0.530	0.683	0.854	1.055	1.310	1.697	2.042	2.457	2.750	3.646
40	0.126	0.255	0.388	0.529	0.681	0.851	1.050	1.303	1.684	2.021	2.423	2.704	3.551
60	0.126	0.254	0.387	0.527	0.679	0.848	1.046	1.296	1.671	2.000	2.390	2.660	3.460
120	0.126	0.254	0.386	0.526	0.677	0.845	1.041	1.289	1.658	1.980	2.358	2.617	3.373
∞	0.126	0.253	0.385	0.524	0.674	0.842	1.036	1.282	1.645	1.960	2.326	2.576	3.291

TABLE V

THE F DISTRIBUTION

For a given percentage p, the p % value F_p of the F distribution is defined by the condition (cf. 8.3)

$$P(F > F_p) = \frac{p}{100}.$$

Thus the probability that F assumes a value exceeding F_p is equal to p %.

Degrees of freedom in denominator n	5 % value F_5 as function of m and n									
	Degrees of freedom in numerator, m									
	1	2	3	4	5	6	8	12	24	∞
1	161.4	199.5	215.7	224.6	230.2	234.0	238.9	243.9	249.0	254.3
2	18.51	19.00	19.16	19.25	19.30	19.33	19.37	19.41	19.45	19.50
3	10.13	9.55	9.28	9.12	9.01	8.94	8.84	8.74	8.64	8.53
4	7.71	6.94	6.59	6.39	6.26	6.16	6.04	5.91	5.77	5.63
5	6.61	5.79	5.41	5.19	5.05	4.95	4.82	4.68	4.53	4.36
6	5.99	5.14	4.76	4.53	4.39	4.28	4.15	4.00	3.84	3.67
7	5.59	4.74	4.35	4.12	3.97	3.87	3.73	3.57	3.41	3.23
8	5.32	4.46	4.07	3.84	3.69	3.58	3.44	3.28	3.12	2.93
9	5.12	4.26	3.86	3.63	3.48	3.37	3.23	3.07	2.90	2.71
10	4.96	4.10	3.71	3.48	3.33	3.22	3.07	2.91	2.74	2.54
11	4.84	3.98	3.59	3.36	3.20	3.09	2.95	2.79	2.61	2.40
12	4.75	3.88	3.49	3.26	3.11	3.00	2.85	2.69	2.50	2.30
13	4.67	3.80	3.41	3.18	3.02	2.92	2.77	2.60	2.42	2.21
14	4.60	3.74	3.34	3.11	2.96	2.85	2.70	2.53	2.35	2.13
15	4.54	3.68	3.29	3.06	2.90	2.79	2.64	2.48	2.29	2.07
16	4.49	3.63	3.24	3.01	2.85	2.74	2.59	2.42	2.24	2.01
17	4.45	3.59	3.20	2.96	2.81	2.70	2.55	2.38	2.19	1.96
18	4.41	3.55	3.16	2.93	2.77	2.66	2.51	2.34	2.15	1.92
19	4.38	3.52	3.13	2.90	2.74	2.63	2.48	2.31	2.11	1.88
20	4.35	3.49	3.10	2.87	2.71	2.60	2.45	2.28	2.08	1.84
21	4.32	3.47	3.07	2.84	2.68	2.57	2.42	2.25	2.05	1.81
22	4.30	3.44	3.05	2.82	2.66	2.55	2.40	2.23	2.03	1.78
23	4.28	3.42	3.03	2.80	2.64	2.53	2.38	2.20	2.00	1.76
24	4.26	3.40	3.01	2.78	2.62	2.51	2.36	2.18	1.98	1.73
25	4.24	3.38	2.99	2.76	2.60	2.49	2.34	2.16	1.96	1.71
26	4.22	3.37	2.98	2.74	2.59	2.47	2.32	2.15	1.95	1.69
27	4.21	3.35	2.96	2.73	2.57	2.46	2.30	2.13	1.93	1.67
28	4.20	3.34	2.95	2.71	2.56	2.44	2.29	2.12	1.91	1.65
29	4.18	3.33	2.93	2.70	2.54	2.43	2.28	2.10	1.90	1.64
30	4.17	3.32	2.92	2.69	2.53	2.42	2.27	2.09	1.89	1.62
40	4.08	3.23	2.84	2.61	2.45	2.34	2.18	2.00	1.79	1.51
60	4.00	3.15	2.76	2.52	2.37	2.25	2.10	1.92	1.70	1.39
120	3.92	3.07	2.68	2.45	2.29	2.17	2.02	1.83	1.61	1.25
∞	3.84	2.99	2.60	2.37	2.21	2.09	1.94	1.75	1.52	1.00

TABLE VI

THE F DISTRIBUTION

For a given percentage p. the $p\%$ value F_p of the F distribution is defined by the condition (cf. 8.3)

$$P(F > F_p) = \frac{p}{100}.$$

Thus the probability that F assumes a value exceeding F_p is equal to $p\%$.

Degrees of free-dom in denom-inator n	1% value F_1 as function of m and n									
	Degress of freedom in numerator, m									
	1	2	3	4	5	6	8	12	24	∞
1	4052	4999	5403	5625	5764	5859	5981	6106	6234	6366
2	98.49	99.01	99.17	99.25	99.30	99.33	99.36	99.42	99.46	99.50
3	34.12	30.81	29.46	28.71	28.24	27.91	27.49	27.05	26.60	26.12
4	21.20	18.00	16.69	15.98	15.52	15.21	14.80	14.37	13.93	13.46
5	16.26	13.27	12.06	11.39	10.97	10.67	10.27	9.89	9.47	9.02
6	13.74	10.92	9.78	9.15	8.75	8.47	8.10	7.72	7.31	6.88
7	12.25	9.55	8.45	7.85	7.46	7.19	6.84	6.47	6.07	5.65
8	11.26	8.65	7.59	7.01	6.63	6.37	6.03	5.67	5.28	4.86
9	10.56	8.02	6.99	6.42	6.06	5.80	5.47	5.11	4.73	4.31
10	10.04	7.56	6.55	5.99	5.64	5.39	5.06	4.71	4.33	3.91
11	9.65	7.20	6.22	5.67	5.32	5.07	4.74	4.40	4.02	3.60
12	9.33	6.93	5.95	5.41	5.06	4.82	4.50	4.16	3.78	3.36
13	9.07	6.70	5.74	5.20	4.86	4.62	4.30	3.96	3.59	3.16
14	8.86	6.51	5.56	5.03	4.69	4.46	4.14	3.80	3.43	3.00
15	8.68	6.36	5.42	4.89	4.56	4.32	4.00	3.67	3.29	2.87
16	8.53	6.23	5.29	4.77	4.44	4.20	3.89	3.55	3.18	2.75
17	8.40	6.11	5.18	4.67	4.34	4.10	3.79	3.45	3.08	2.65
18	8.28	6.01	5.09	4.58	4.25	4.01	3.71	3.37	3.00	2.57
19	8.18	5.93	5.01	4.50	4.17	3.94	3.63	3.30	2.92	2.49
20	8.10	5.85	4.94	4.43	4.10	3.87	3.56	3.23	2.86	2.42
21	8.02	5.78	4.87	4.37	4.04	3.81	3.51	3.17	2.80	2.36
22	7.94	5.72	4.82	4.31	3.99	3.76	3.45	3.12	2.75	2.31
23	7.88	5.66	4.76	4.26	3.94	3.71	3.41	3.07	2.70	2.26
24	7.82	5.61	4.72	4.22	3.90	3.67	3.36	3.03	2.66	2.21
25	7.77	5.57	4.68	4.18	3.86	3.63	3.32	2.99	2.62	2.17
26	7.72	5.53	4.64	4.14	3.82	3.59	3.29	2.96	2.58	2.13
27	7.68	5.49	4.60	4.11	3.78	3.56	3.26	2.93	2.55	2.10
28	7.64	5.45	4.57	4.07	3.75	3.53	3.23	2.90	2.52	2.06
29	7.60	5.42	4.54	4.04	3.73	3.50	3.20	2.87	2.49	2.03
30	7.56	5.39	4.51	4.02	3.70	3.47	3.17	2.84	2.47	2.01
40	7.31	5.18	4.31	3.83	3.51	3.29	2.99	2.66	2.29	1.80
60	7.08	4.98	4.13	3.65	3.34	3.12	2.82	2.50	2.12	1.60
120	6.85	4.79	3.95	3.48	3.17	2.96	2.66	2.34	1.95	1.38
∞	6.64	4.60	3.78	3.32	3.02	2.80	2.51	2.18	1.79	1.00

INDEX

279

280